PRACTICAL PRINCIPLES OF ION EXCHANGE WATER TREATMENT

By
Dean L. Owens

TALL OAKS PUBLISHING, INC.　　　　VOORHEES, NJ

Library of Congress Catalog Card Number: 85-51869

FIRST PRINTING

COPYRIGHT 1985 by TALL OAKS PUBLISHING, INC.   ALL RIGHTS RESERVED
Neither this book, nor any part of it, may be reproduced or transmitted in any form or by any means, electronic or mechanical, including photocopying, microfilming, and recording, or by any information storage and retrieval system, without permission in writing from the publisher.

TALL OAKS PUBLISHING, INC.
1507 Evesham Road
Voorhees, NJ 08043

PRINTED IN THE UNITED STATES OF AMERICA

DEDICATION

The author dedicates this book to the memory of

Dr. Fredrick J. Myers

whose encouragement and advise enhanced his career.

FOREWORD

My writing the FOREWORD to this work has indeed been a pleasure since I have known and been associated with the author throughout his entire professional career.

Although a host of books have been written on the treatment of water by ion exchange, none have been written for the technicians and operators who have but little formal education and training in chemistry and engineering. This book, "The Practical Principles of Ion Exchange Water Treatment," has been written for this audience. The successful operation of *all* ion exchange installations depends upon the training and understanding of the operators. This book should be a "must" for this group.

Dean Owens should be praised and complimented for his efforts in fulfilling the gap that has been present in the literature on water treatment and ion exchange technology.

Robert Kunin

PREFACE

My first introduction to ion exchange technology occurred in 1941, while working on a summer job to support my education at the University of Wisconsin. It was not until the fall of 1946 when I once again became involved with this interesting technology, after being employed as a chemist by Rohm and Haas Co. My activities connected with the sales development of ion exchange resins continued throughout my career with Rohm and Haas Co. to the time of my early retirement in 1976.

My fascination with ion exchange technology and its applications for a wide variety of uses from A(tomic bombs) to Z(irconium) lead to my desire to solicit consulting work in this field. Fortunately, the number of people at that time who had in-depth familiarity with ion exchange technology were not great, thus providing me an opportunity to have achieved success in my consulting career and the time to write this book.

It is my hope that this book may be of help in improving the understanding of ion exchange for those persons with little technical background on the subject, who are involved with the operation, supervision or management of ion exchange systems. The major emphasis of this book is directed to water treatment, although the technology may be applicable to other areas of ion exchange.

This book has been written at the suggestion and encouragement of many of my associates, friends and clients. Its purpose is to aid in the training of personnel who may not have sufficient technical training or background to be able to understand the many excellent technical books that have been written on the subject. I would hope that this book will only be the beginning point in education on the subject of ion exchange and will encourage the reader to go on to further develop their knowledge by reading those books listed in this book's addendum, as well as other excellent discourses on the subject.

I would like to express my gratitude to Dr. Robert Kunin, who has always been an invaluable source of knowledge in ion exchange technology, for his assistance and encouragement in the preparation of this book. I am also very indebted to Mr. William C. Burns who spent generously of his time in assisting with the preparation of this book and contributing to my knowledge of ion exchange equipment and systems. I would also like to acknowledge and thank Dr. Harold E. Weaver for his assistance and guidance throughout much of my career in sales of ion exchange resins which eventually made possible the opportunity to write this book.

CONTENTS

Dedication		*iii*
Foreword		*v*
Preface		*vii*
Chapter 1	Introduction	1
Chapter 2	Ion Exchange Terminology	3
Chapter 3	The Importance of Water Analysis to Ion Exchange	9
Chapter 4	Ion Exchange Resins and Their Reactions	21
Chapter 5	Equipment for Ion Exchange	33
Chapter 6	Ion Exchange Cycles	57
Chapter 7	Ion Exchange Systems	99
Chapter 8	Design and Sizing of Ion Exchange Systems	123
Chapter 9	Troubleshooting	133
Chapter 10	Waste	145
	In Closing	149
Appendix		
	Glossary	*151*
	Tables and Conversion Factors	*159*
	Supplemental Reading References	*179*
	Subject Index	*181*

CHAPTER 1

INTRODUCTION

Although ion exchange reactions were recognized in the 19th century, some scholars point out much earlier accounts of occurrences that can only be explained as ion exchange reactions: those found in the Bible in the Book of Exodus 15:22-25 and the Book of Numbers 19:5-17, and in Sir Francis Bacon's *New Atlantis* and *Sylva Sylvarum*.

These accounts mention, respectively: bitter water made sweet by a tree cast into the water (probably by oxidized cellulose in the tree removing ions from the water); purifying water with mixed vegetables and bone char; pools that strain water out of salt; and salt water being made fresh by passing through earth in twenty vessels.

There is much for us to learn from the past—both distant and recent—if we can only recognize what it is that actually happened and how to apply it to our present problems. To do this we must know or learn the lessons by understanding the words used, and then the combination of words will make some sense so we can apply the knowledge gained. In short, we must learn the vocabulary before we can understand the language.

Ion exchange technology has its own vocabulary and I believe that much of the frustration people have in learning the subject, unless they have the benefit of previous education in chemistry, engineering, hydraulics, kinetics, and other related sciences, is difficulty with the terms. In many other subjects, such as medicine or even the newer computer technology, the terms used have to be defined and become familiar terms before we can know "what they are talking about."

The total number of ion exchange water treatment systems installed and operating has been steadily and rapidly increasing since their first introduction in the United States (about 1912) for water softening. With the development of synthetic cation and anion ion exchange resins in 1935, the capability of partial deionization was realized, but not until the 1940s, when strongly basic ion exchange resins were developed, was it possible to completely deionize water.

The further development of newer improved and modified ion exchange resins, along with the improved and more sophisticated equipment and systems in which they are used, has led to the need for trained operating personnel, as well as management and supervisory persons, who are knowledgeable about ion exchange technology.

Understanding the design of the equipment and the instructions for its operation will aid in recognition of malfunctions and corrective procedures to be used. The variations in operation and limits of the equipment and systems will also be important factors in controlling the performance from quality, quantity, and efficiency standpoints. The potential contaminants and foulants and their effect on the system's performance must be understood so that corrective or maintenance actions can be taken.

Many of the operational variables and the causes and effects influencing ion exchange systems are learned by experience. However, when the operating and supervisory personnel have a better understanding of the ion exchange resins and what goes on within the equipment, then better and more efficient operation will be achieved with less downtime and crises.

Ion exchange technology deals with many variables including the ion exchange resins, regeneration chemicals, concentration of dissolved solids in the feedwater, temperature, flow rates, contact time, equipment configuration, distribution systems, and many more. All this could discourage anyone into thinking that a good workable understanding of the subject could not be achieved without in-depth technical background. Yet there are some basics and "rules of thumb" that will allow anyone to reach that goal. Once the basics are learned, then the application of this information to the operation of ion exchange equipment can be made.

It is unfortunate, but true, that water treatment has been delegated to a very unimportant and neglected position in many plants' operations and budgets. In most instances, however, if the water treatment system malfunctions or shuts down, the whole plant will have to shut down. It also should be recognized that there are certainly few other plant systems where a higher demand on quality or purity exists than is routinely produced by ion exchange water treatment systems. In many of today's high quality water treatment systems the control specifications are written requiring dissolved solids and other impurities of no greater than 10 parts per billion (ppb). Written in percent purity this would be 99.999999% pure! This quality is produced in amounts of over 1,000 gallons per minute (over 4 tons per minute) routinely. It is difficult to name many other materials, processed or manufactured, that can make that claim in both purity and volume. It is truly a remarkable feat and one that is achieved only by great care and close attention to detail by operating personnel.

CHAPTER 2

ION EXCHANGE TERMINOLOGY

Many authors of books on the subject of ion exchange start with the assumption that the readers are familiar with the terminology to be used in the discourse or discussion, but I will start with the assumption that even though many readers may have heard of the terms (and many may understand some or all of the terms), some of them are not clearly understood. In order for all to start at the same point, this chapter will deal with definitions of words or terms that are most frequently used in ion exchange and water treatment and hopefully will develop the base for understanding the subject matter in the following chapters.

Most of the terms and unfamiliar words used in ion exchange water treatment will be defined as they appear in the text of this book. A more complete alphabetical listing for later reference is given in the Glossary of this book.

Water
Since water is the raw material and also the end product of ion exchange processing, then it might be in order to look in greater detail at the material we are working with.

Most people would say that any darn fool knows what water is and many people would agree that H_2O is a well-known chemical symbol for water. However, many people utilize the word water alone to describe oceans, rivers, lakes, wells, rain, and drinking, softened, distilled, and even deionized water. It is also quite obvious that the major differences between those types of water are the amount and kinds of "things" or contaminants that are in those waters. Besides whales, fish, plant life, and microorganisms present (among other "things") in ocean or sea water there is also a greater amount of salts, mainly sodium chloride, than in most other waters.

Generally the ocean salinity or salt content (although salt content does vary in many seas around the world) is around 35,000 parts per million (ppm), or 3.5% salt. In the fresh (not salty) lakes and rivers, the dissolved solids (including sodium chloride, but also other salts or minerals such as calcium and magnesium bicarbonates, sulfates and chlorides, and others) run in the range of about 100 to 500 ppm, or 0.01 to 0.05%. Distilled water, with careful control, can be obtained with less than 1 ppm dissolved solids, or 0.0001%. Deionized water can be produced with less than 10 parts per *billion* (ppb) dissolved solids or 0.000001%! At this point what remains is almost pure water or H_2O.

H_2O is, of course, two atoms of hydrogen with one atom of oxygen, a very stable molecule. However, in ion exchange work it is desirable to look at the water molecule existing as H^+OH^- rather than H_2O. Indeed water does exist in this fashion, but in very small quantities as compared to the very stable and only slightly dissociated (the process of ionization of an electrolyte) H_2O as illustrated in this fashion:

$$H_2O \rightleftharpoons H^+ + OH^-$$

with the size of the arrows showing that the equilibrium (the status when a reaction reaches a state of balance) is favoring greatly the H_2O form. The actual amount of H^+ and OH^- ions that exist at normal temperatures is exceedingly small, about one part per one hundred trillion of H_2O. The major point being made is that water itself can dissociate to form ions of the <u>cation</u> hydrogen (H^+) and the <u>anion</u> hydroxyl (OH^-) similar to the way salts, acids, and bases dissociate when dissolved in water to form their corresponding cations and anions.

Ions

Salt, by chemical definition, is what is formed as a result of the reaction of an acid and a base, such as hydrochloric (muriatic acid or HCl) acid reacting with sodium hydroxide (caustic or NaOH) to form salt (NaCl) and water (HOH or H_2O).

The chemical reaction would be illustrated:

$$HCl + NaOH \rightarrow NaCl + H_2O$$

Similarly, other salts could be formed by sodium hydroxide reacting with sulfuric acid (H_2SO_4) to form sodium sulfate (Na_2SO_4); or calcium hydroxide ($Ca(OH)_2$) reacting with hydrochloric acid (HCl) to form calcium chloride ($CaCl_2$). Sodium chloride (table salt) in its crystalline form when added to water dissolves (up to its solubility limit of about 26%) and cannot be seen in the water. The dissolving process causes the sodium to dissociate from the chloride and now the positively charged (+) sodium <u>ion</u> is free to move in the water more independently from the negatively charged (-) chloride <u>ion</u> than when they were in the crystalline solid state. This could be illustrated as shown below

$$\text{(crystalline solid) NaCl} \xrightarrow{\text{water}} Na^+ \text{ and } Cl^- \text{ (dissolved in water)}$$

Chapter 2

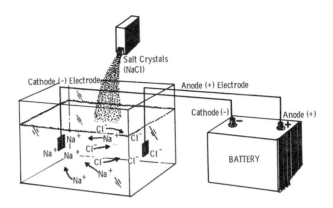

Figure 2-1. Dissolution of salt into ions and migration to opposite charged electrode in chemical cell.

The same would happen with other salts as they dissolve into a water solution, becoming <u>positive ions</u> and <u>negative ions</u>.

Anions and Cations

Now let us make up a salt (NaCl) solution into which we will immerse two electrodes and connect the electrodes to a battery, as shown in Fig 2-1.

The old but true saying applies here that opposites attract and the <u>positively charged</u> sodium <u>ions</u> will go to the negatively charged electrode or <u>cathode</u> while the <u>negatively charged</u> chloride <u>ions</u> go to the positively charged electrode or anode. The <u>ions</u> attracted to the <u>cathode</u> are called <u>cations</u> and <u>ions</u> attracted to the <u>anode</u> are called <u>anions</u>. All positive charged ions are cations and all negative charged ions are anions.

There are numerous types of ions that can be present in water, with one difference being in the number of charges that are on the ions. With sodium (Na^+) there is only one positive charge on the ion, while with the calcium (Ca^{+2}) and magnesium (Mg^{+2}) there are two positive charges, and aluminum (Al^{+3}) has three positive charges, as examples. The more charges on the cation, the more strongly it will be attracted to the cathode or negatively charged electrode. In the case of anions the same is true, with chloride (Cl^-) having one negative charge, sulfate (SO_4^{-2}) two, and phosphate (PO_4^{-3}) three, as examples. The attraction of the anions toward the anode or positively charged electrode is stronger for those with the largest number of charges.

In ion exchange water treatment we are primarily concerned with the cations and anions shown in Table 2-A.

TABLE 2-A
Typical Cations and Anions Present in Water

Major

Cations	Anions
Calcium (Ca^{+2})	Bicarbonate (HCO_3^-)
Magnesium (Mg^{+2})	Sulfate (SO_4^{-2})
Sodium (Na^+)	Chlorides (Cl^-)
	Silica (SiO_2^{-2})*

* Silica is present in many forms, as a very weak acid or polymers. More details will be given in later discussions.

Minor

Cations	Anions
Potassium (K^+)	Carbonate (CO_3^{-2})
Barium (Ba^{+2})	Nitrate (NO_3^-)
Iron (Fe^{+2} or Fe^{+3})	Phosphate (PO_4^{-3})
Manganese (Mn^{+2} or Mn^{+3})	Sulfide (S^{-2})
Aluminum (Al^{+3})	Fluoride (F^-)
Ammonium (NH_4^+)	Carbon Dioxide (CO_2)*
Strontium (Sr^{+2})	

* Carbon dioxide dissolves in water to become the weak acid, carbonic acid. More details later.

The source of these ionic "contaminants" of water depends on where the water comes from and the environment the water was exposed to, as well as any pretreatment it received before the ion exchange treatment. With increasing probabilities of industrial, agricultural, human, and other environmental "contaminants" added to the natural sources, the minor or trace contaminants become numerous. Further discussion of these potential problems will be covered in Chapter 9, "Troubleshooting."

Ion Exchange

Once the definition of ions is understood, the term "ion exchange" is rather simple, since it means that one ion in solution is exchanged or replaced by another. If both ions were in a water solution there would be no exchange, since they both would be free to move around in the water solution and nothing would be removed or replaced. It is only when one ion is attached to an insoluble solid and the other is in the solution that an exchange or replacement can be carried out. The fact should be pointed out that an ion exchange can only be carried out with ions of the same charge, that is, positive ions for positive ions, or negative ions for negative ions.

The earliest reported and verified studies on ion exchange took place in the mid-1850s. The work of Thompson and Way showed that when soils were treated with ammonium sulfate or ammonium carbonate, most of the ammonium ion (NH_4^+) was absorbed and calcium ion (Ca^{+2}) was re-

Chapter 2

leased. This ion exchange reaction explains how ammonia fertilization of soils is accomplished to allow plant growth. The aluminum silicates present in the soils were found to be the solid, insoluble substances responsible for holding the ion to be exchanged for an ion in the water solution. This was the first recognized ion exchange mineral or substance.

Some other terms used with great frequency in ion exchange discussions, and not explained fully in the Glossary, will be introduced here and will be discussed further in Chapter 3, "Water Analysis," which will show how they can be expressed in quantity for ion exchange calculations. These terms are:

Total Dissolved Solids (TDS)
This term is frequently misunderstood, probably because of the various similar terms such as "total solids, total suspended solids, and dissolved solids." The term "total solids" is the sum of suspended and dissolved solids, while "total suspended solids" is the amount of solids which are suspended and not in true solution. To determine total solids, a measured volume of water is evaporated under controlled temperature and the weight of the residue is calculated to give results in parts per million (ppm). Another sample of the same water is filtered to remove all suspended solids and then the same evaporation procedure is run on this measured volume of water, as above, and the residue weighed to give a value for dissolved solids in ppm, with the difference between the two values being total suspended solids. Total dissolved solids and dissolved solids are used interchangeably as an expression of the soluble constituents in the water. In the reverse osmosis industry, however, total dissolved solids (TDS) is defined as the total sum of all cations and anions in ppm as ions*, while in the ion exchange industry it is the sum of either cations or anions in ppm as $CaCO_3$* or epm*, but not the sum of cations and anions.

Hardness
The hardness of water is the combined amount of calcium (Ca^{+2}) and magnesium (Mg^{+2}) present in the water in grains per gallon (gr/gal)* or in ppm as $CaCO_3$.* Other divalent or trivalent cations such as iron, barium, or strontium should also be included if present.

The hardness of water is the cause of scale formation in pipes, teakettles, boilers, and the familiar bathtub ring. It is likely that the term hardness originated from the formation of a hard scale formed by the calcium and magnesium in the pipes and boilers. When the hardness has been removed from the water and replaced with sodium, the water is called soft water.

Free Mineral Acids (FMA) and Total Mineral Anions (TMA)
Acids coming from mineral sources such as chlorides, sulfates, nitrates, etc., are strong acids and thus the name, mineral acids. By comparison, acids from

organic sources such as carbonic acid (H_2CO_3 or carbon dioxide dissolved in water), acetic acid (vinegar), and citric acid (lemon juice) are generally weak acids. The FMA or TMA of normal water analysis therefore are the sum of the chloride and sulfate concentrations (although fluorides, nitrates, and others would be included, if present) when they are expressed in epm* or ppm as $CaCO_3$*. The FMA of the water coming from a strongly acidic cation resin in the hydrogen form may be less than the TMA of the water by the amount of the sodium leakage.

Alkalinity

The alkalinity of a water is determined by the amount of bicarbonate (HCO_3^-), carbonate (CO_3^{-2}) or hydrate (hydroxyl OH^-) present in the water. Most raw waters contain only bicarbonate alkalinity, while treated waters may contain bicarbonate and carbonate or carbonate and hydrate alkalinity.

pH

In simplest terms, pH is a number system that relates to the acidity, neutrality, or basicity (alkalinity) of a water with numbers from 0 (most acidic) to 7 (neutral) to 14 (most basic). More precisely, it is a measure of the hydrogen ion (H^+) concentration of the water.

Pure water (without other ionized substances or contaminants) has a hydrogen ion concentration of 0.0000001 grams of hydrogen ion in each liter of water at 25°C. This number can be expressed as 1×10^{-7} or on the pH scale as 7 or neutral. Each unit of pH below 7, such as 6, represents a 10-fold increase in hydrogen ion concentration from the unit above it. The reverse is also true. Each pH unit above 7 is a 10-fold decrease in hydrogen ion concentration from that of the previous unit.

Free CO_2 (Carbon Dioxide)

This term is to distinguish carbon dioxide gas dissolved in water as differentiated from combined carbon dioxide in the water in the form of bicarbonate or carbonate ions.

---

* These terms will be discussed in detail in Chapter 3 on water analysis, showing the difference in expressing concentration on a weight or on an equivalent weight basis.

CHAPTER 3

THE IMPORTANCE OF WATER ANALYSIS TO ION EXCHANGE

There are many texts and manuals available on the methods used in running a water analysis, as well as standard methods of the American Society of Testing and Materials (ASTM) and American Water Works Association (AWWA). Thus this subject need not be covered in any detail other than to define the terminology used to report the results of the analysis and to show how these terms can be used and converted to determine capacities of ion exchange resins for any particular water.

For the majority of applications of ion exchange resin water treatment, it is necessary to know what ions and how much or what concentrations of each are present. For applications such as water softening, generally just knowing the amount of hardness present (calcium, magnesium, and if present, iron, manganese, strontium, and barium) is sufficient. However, where the softening of water high in sodium chloride content is desired, then knowing the total dissolved solids and sometimes having a complete analysis are necessary in order to determine the best softening approach and to determine the capacity of the cation exchange resin to be used.

For deionization, the complete water analysis must be run and it is best to include analyses for some of the minor constituents, since some of these may have an influence on the regeneration techniques to be used and can affect the capacities that can be obtained. Of particular concern are the cations and the anions with trivalent charges, since these are not as readily removed in regeneration and can build up, reducing capacities for removal of other ions.

Barium and strontium, although only divalent, form extremely insoluble sulfates and therefore sulfuric acid can not be used as a cation resin regenerant when these ions are present. Fortunately, these types of ions are not frequently present in water supplies, but it is wise to have analyses run for some of these minor constituents on waters where there has been no previous experience with ion exchange systems.

A word or two might be desirable with regard to caution on obtaining an analysis of water. The ability of some technicians to obtain good reproducible analytical results can be questioned. Based on experience, test results have been observed, from different laboratories on the same water sample, to vary by greater than 10% between each other.

It is also very desirable to have more than one sample of water tested over periods of time so as to cover seasonal variations that can be expected to occur, with surface waters in particular. In cities or municipal water supply areas there may be blending of waters from various sources which can vary on a day-to-day basis and in order to get a good profile of the maximum and minimum water analyses, samples should be taken over an extended period of time. Water analyses can usually be obtained from the water utility supplying your facility, but sometimes these are not complete, covering all the ions you are interested in for ion exchange capacity calculations. Also, some analyses are on the raw water supplies and others may be given for the treated water after flocculation, clarification, filtration, and chlorination, which can significantly change the chemical composition of the water delivered to your facility.

A variation of 10% in water composition may not seem great, but if your deionizer is rated to produce 100,000 gallons per service cycle and only produces 90,000 gallons, there might be cause for concern unless you do take into consideration the effect of changes in the raw water supply. The changes in the balance of ions in the raw water, even with the total dissolved solids remaining the same, can also affect the capacity that can be expected from the ion exchange system, and this will be covered in Chapters 6 and 9.

Methods of Reporting Water Analyses

Four methods used to report water analyses are given here.

Method 1: ppm as ion or mg/l (milligrams per liter). In determining the amounts of each ion in a sample of water, a specific chemical test is run and reported in terms of the weight of that substance in a given weight of water. Usually this is given in parts per million (ppm) or milligrams per liter (mg/l) and both terms are interchangeable, as 1 ppm means 1 part (whether it be pound, ton, gram, ounce, milligram, or other unit of weight) in 1,000,000 parts of the same unit of water. A liter is an expression of volume, but fortunately 1 liter weighs 1,000 grams, so 1,000 grams equals 1,000,000 milligrams, and therefore the metric expression is the same as the ppm.

As an example, a water analysis, "In ppm as ion," is given in Table 3-A.

The interesting aspect of the analysis reported above in "ppm as ion" is that you do have the actual weight reported for each ion in 1 million parts by weight of water, which would be 125 parts of total cations and 303 parts of total anions for a total of 428 parts by weight of total ions.

If we took 1,000,000 pounds of water plus 428 pounds of the ions in the Table 3-A analysis, we would have 1,000,428 pounds of this water. If we evaporated off the 1,000,000 pounds of water, we would be left with a 428-

Chapter 3

TABLE 3-A
ppm as Ion

Cations		Anions	
Calcium as Ca	40	Bicarbonate as HCO_3	100
Magnesium as Mg	25	Sulfate as SO_4	50
Sodium as Na	60	Chlorides as Cl	141
Total	125	Subtotal	291
		SiO_2	12
		Total	303

pound residue. This process of evaporating off the water is just what is done to determine a figure for the total dissolved solids by gravimetric (measure by weight) analysis. Sample sizes used in the laboratory might be 50 to 1,000 milliliters of water. In the normal laboratory process of drying, a temperature of 103°C (or 217.4°F) would tend to volatilize the free carbon dioxide that might be present. This would reduce the reported total dissolved solids to less than that determined by other analytical techniques, and could be the reason for small variations in reported results.

Although we have a value for total weight of ions in the water by adding up the weight of all the ions, it is like adding up all the different kinds of fruit in a basket that may include apples, pears, peaches, oranges, grapefruit, etc., and coming up with so many pounds of total fruit.

Carrying the analogy further, if each type of fruit has a different price (as the ions have positive or negative charges of varying amounts), and we would like to know how much fruit we can put in a shopping basket (as compared to the capacity of cation and anion resins to take up ions), it is necessary to make some distinction between the fruit in terms of dollars per pound each might cost.

Let us say oranges are 25 cents per pound and grapefruit are 50 cents per pound (compared to 1 positive charge on sodium ion and 2 positive charges on calcium). If we have 4 dollars to spend, we can take into our shopping basket 8 pounds of grapefruit or 16 pounds of oranges, or any combination such as 4 pounds of oranges and 6 pounds of grapefruit. It can be seen that the common denominator in the fruit illustration would be cost per pound to fill the shopping basket, while with ions it would be charges per ion weight to determine how many may be put on a resin that has a fixed capacity.

Since the weight of the sodium cation is different from that of the calcium cation (as the weight of the orange may be different from that of the grape-fruit) we must come up with a means of making them equivalent in the amount of capacity they will take up in the ion exchange resin beads (as the amounts of fruit we can put in the shopping basket is determined by the dollars per pound they cost as related to the money we have to buy fruit.) This is done by dividing the weight of the ion by its charge. Sodium has one charge, so the equivalent weight is its ion or molecular weight. Calcium and magnesium have two charges, so their weight would be their molecular weight divided by two. The same type of reasoning can be used with anions.

Now with each ion having an equivalent weight we can add the total equivalent weight of cations and come up with the total of positively charged ions. Likewise, we can add up the total equivalent weight of anions and come up with the total of negatively charged ions. With a water of an approximate neutral pH of 7 the sum of cations should equal the sum of anions.

These charges of the ions, in chemical terms, are called valences. One charge (either positive or negative) is called monovalent; two charges, divalent; three charges, trivalent, etc. Therefore, the term "equivalent weight" means on an equal valence basis of weight (ionic or molecular).

From the above discussions we can come up with the following:

$$\text{Equivalent} = \text{equal valence}$$

$$\text{Equivalent weight} = \frac{\text{ionic or molecular weight}}{\text{valence}}$$

For example, sodium with a molecular weight of 23 and a valence of 1 would have an equivalent weight of $23 \div 1 = 23$, while calcium with a molecular weight of 40.1 and a valence of 2 would have an equivalent weight of $40.1 \div 2 = 20$.

TABLE 3-B
Conversion from ppm to epm

Cations	ppm/eq-wt	epm
Ca	$40 \div 20$	2.00
Mg	$25 \div 12.2$	2.05
Na	$60 \div 23$	2.61
Anions		
HCO_3	$100 \div 61$	1.64
SO_4	$50 \div 48$	1.04
Cl	$141 \div 35.5$	3.97
SiO_2	$12 \div 60$ *	0.20

*SiO_2 is considered to have an equivalent weight of 60 since it exists as a monovalent ion $HSiO_2$

Chapter 3

Method 2: Epm or meq/l (milliequivalents per liter). To obtain the epm from the water analysis expressed in ppm as ion shown in Table 3-A, it is necessary to divide the unit weight (ppm as ion) of each substance by the equivalent weight as shown in Table 3-B.

$$\text{Equivalence per million} = \frac{\text{ppm as ion}}{\text{Equivalent weight}}$$

The water analysis in epm is shown in Table 3-C.

TABLE 3-C
EPM

Cations		Anions	
Ca	2.00	HCO_3	1.64
Mg	2.05	SO_4	1.04
Na	2.61	Cl	3.97
Total*	6.66	Subtotal*	6.65
		SiO_2	0.20
		Total	6.85

*These values should be the same except for rounding off of the third decimal point.

Now that all the substances are converted to an equivalent basis, it is possible to add the cations to get total cations and add anions for total anions. It can be seen that the sum of the cations and the sum of the anions (disregarding silica) come to approximately the same amount, which should be the case with water that would have a nearly neutral pH. This is also a good check on whether the water analysis is a good one, since the cations and the anions should be balanced to have a neutral water and to have electroneutrality.

Method 3: ppm as $CaCO_3$. The third way to express the water analysis is the one most commonly used in calculations to determine the capacity of ion exchange resin systems in English terms. The conversion to ppm as $CaCO_3$, from either ppm as ion or epm, is easily done with the understanding developed in the previous sections that the expressions of concentration of each substance are on a weight basis and are convertible by taking into account that each substance or ion has a different ionic or molecular weight and can have a different valence.

Since $CaCO_3$ is a combination of Ca^{+2} and CO_3^{-2}, both divalent, the equivalent weight would be obtained by dividing the molecular weight of 100.08 by 2 to give a value of 50.04. Now all that is necessary to obtain the analysis in ppm as $CaCO_3$ is to multiply the epm value of each ion in Table 3-C analysis by 50 to obtain the analysis as shown in Table 3-D.

To convert to ppm as $CaCO_3$ from ppm as ion, simply multiply the ppm as ion by the ratio of the equivalent weight of $CaCO_3$ to the equivalent weight of the specific ion being converted.

Or: ppm as $CaCO_3$ =   ppm as ion $\times \dfrac{50}{\text{Equivalent weight of ion}}$

For example, magnesium equivalent weight is 12.2, so: $50 \div 12.2 = 4.12$. Multiply 4.12 by the ppm as ion of magnesium shown in Table 3-A, or:

$$4.12 \times 25 = 103 \text{ ppm as } CaCO_3.$$

For anions the same procedure would apply. For other ion conversion factors, see Table 3-F.

The sample water analysis in ppm as $CaCO_3$ would then be as shown in Table 3-D.

Method 4: grains per gallon as $CaCO_3$. Expressing the water analysis in grains per gallon as $CaCO_3$ makes it convenient when calculating the capacity of the ion exchange resins (in English terms) to the number of gallons of water that can be treated per cubic foot of resin.

The term grain originally developed from the fact that it required 7,000 average grains of dry wheat to equal one pound. When looking at a water analysis expressed in grains per gallon, it might be helpful as a visual aid to the reader to imagine that number of grains of wheat in a gallon of water as representing the amount of ions in the water. However, it also should be realized that one grain of wheat is very light and weighs only 0.065 grams or 0.0023 ounces.

To convert from ppm as $CaCO_3$ to grains per gallon as $CaCO_3$ we can start with:

$$1 \text{ grain as } CaCO_3 = \dfrac{1}{7{,}000 \text{ grains/pound}} = 0.0001428 \text{ pounds}$$

Since 1 gallon weighs 8.33 pounds, then:

$$1 \text{ grain per gallon} = \dfrac{0.0001428 \times 1{,}000{,}000}{8.33} = 17.1 \text{ as } CaCO_3$$

TABLE 3-D
ppm as CaCO$_3$

Cations		Anions	
Ca	100	HCO$_3$	82
Mg	103	SO$_4$	52
Na	130	Cl	199
Total	333	Subtotal	333
		SiO$_2$	10
		Total	343

TABLE 3-E
Grains per Gallon as CaCO$_3$

Cations		Anions	
Ca	5.82	HCO$_3$	4.80
Mg	6.02	SO$_4$	3.04
Na	7.60	Cl	11.64
Total	19.47	Subtotal	19.48
		SiO$_2$	0.58
		Total	20.06

To change the analysis in Table 3-D to grains per gallon as CaCO$_3$, divide the ppm as CaCO$_3$ by 17.1 to obtain the results given in Table 3-E.

Other Water Analysis Terminology

There are other reported values sometimes given in water analyses, or which may be determined from the analyses, which are important for the calculation of the quantity and quality of water that can be produced by ion exchange resins. Some of these have been defined in the Glossary section of the Appendix, but how they are obtained and used will be discussed further in this section.

Alkalinity. In many waters the alkalinity is the amount of bicarbonate present and the percent of alkalinity is the amount of bicarbonate divided by the total anions times 100, from the water analysis expressed in epm, ppm as CaCO$_3$, or grains per gallon as CaCO$_3$. Where carbonate or hydroxide alkalinity are present, then the alkalinity will be the sum of the amounts of each present, or total alkalinity divided by the total anions times 100.

The higher the percent alkalinity, the higher the capacity and the lower the leakage of sodium will be from the cation resins operating in the hydrogen form.

TABLE 3-F
Water Analysis Conversion Factors

Substance	Atomic or Molecular Weight	Equivalent Weight	Substance to $CaCO_3$ Equivalent
Aluminum	27.0	9.0	5.56
Ammonia	17.0	17.0	2.94
Ammonium	18.0	18.0	2.78
Barium	137.4	68.7	0.73
Bicarbonate	61.0	61.0	0.82
Calcium	40.1	20.0	2.50
Carbonate*	60.0	60.0	0.83
CO_2*	44.0	44.0	1.14
Chloride	35.5	35.5	1.41
Iron Fe^{+2}	55.8	27.9	1.79
Iron Fe^{+3}	55.8	18.6	2.69
Magnesium	24.3	12.2	4.12
Nitrate	62.0	62.0	0.81
Phosphate	95.0	31.7	1.58
Potassium	39.1	39.1	1.28
Silica*	60.1	60.1	0.83
Sodium	23.0	23.0	2.17
Strontium	87.6	43.8	1.14
Sulfide	32.1	16.0	3.13
Sulfate	95.1	48.0	1.04

*Carbonate, carbon dioxide and silica react as monovalent ions in ion exchange.

Free mineral acids (FMA) or total mineral acidity (TMA). In essence, the TMA of a water is the anions in the water analysis which are not alkalinity or silica. The FMA of a water is the amount of TMA converted to the acid form by hydrogen ion exchange for the cations associated with the sulfates and chlorides (or fluorides and nitrates, if present) and is only less than the TMA by the amount of sodium ion not exchanged for hydrogen (or sodium leakage).

The higher the percent TMA of total anions in the water analysis (conversely to the alkalinity), the lower the capacity and the higher the leakage of sodium will be for the cation resins operating in the hydrogen form.

Total dissolved solids (TDS). Although conductivity devices are used in obtaining a measure of the TDS of waters, and they are good for monitoring changes in TDS, they are not accurate enough to be used in ion exchange

Chapter 3

calculations to determine capacities or leakage of anion and cation exchange resins. A complete water analysis is the best means of obtaining values of TDS to be used in ion exchange resin calculations.

TDS for ion exchange calculations is the sum of the cations *or* the sum of the anions, depending on which resins are being considered. Using the expression of the water analysis in epm, ppm as $CaCO_3$, or grains per gallon as $CaCO_3$ is satisfactory for figuring percentages and total cations or anions for capacity or leakage values for both cation and anion resins.

pH. Most natural waters will have pH in the range of 6 to 8; however, waters with high amounts of dissolved carbon dioxide can be on the low side near 5, with the formation of carbonic acid. Alkaline waters above pH 8 are seldom found except where some lime pretreatment has been used. When the pH is lower than 7 in a water analysis, the anions can be expected to be slightly higher than the cations. Conversely, if the pH is higher than 7, the cations can be expected to be slightly higher than the anions. If the pH is very nearly neutral there should be a balance of anions and cations in the water analysis (neglecting the silica, which has little effect on the pH. However, the silica must be added to the anions in figuring the total anions for removal by strongly basic anion resins). The pH therefore can be used as a check on the accuracy of the water analysis.

Chlorine. Most municipal treated waters are chlorinated to give a safe and sterile water supply for potable or drinking water purposes. There is a residual free available chlorine of approximately 1 mg/l or 1 ppm in most municipal distribution systems, but in some the residual may be higher or lower, depending on the distance of the supply pipe lines from the chlorination point. Chlorine in solution forms hypochlorous acid. The hypochlorite ion, which is a strong oxidizing agent used to destroy harmful bacteria in the water, can also be destructive to ion exchange resins. Pretreatment for removal of chlorine and its derivatives with sulfur dioxide, sodium sulfite, or carbon may be desirable for protection of the ion exchange resins.

Free CO_2 (carbon dioxide). There may be free carbon dioxide in both raw water and treated waters. Free carbon dioxide is considered to react as a monovalent anion, as it will covert to carbonic acid (H_2CO_3) in water and exchange on the ion exchange resin as bicarbonate (HCO_3^-) and therefore the equivalent weight should be calculated accordingly.

Turbidity. Turbidity in waters is due to finely divided suspended matter, of clay, silt, and/or organic composition. Measurement of turbidity is done by comparing the optical obstruction of light rays passed through a sample of water as compared with a standard turbidity scale. Waters with Jackson Turbidity Units (JTU or JU) over 2 to 10 (depending on the frequency of backwash or length of service run) are usually considered to be a potential problem for filtering out on the ion exchange resin beds. These waters are likely to cause pressure drop, channeling, or capacity problems and therefore should be pretreated by coagulation and filtration.

Formazin Turbidity Units (FTU) and Nephelometric Turbidity Units (NTU) are two other methods used to express amounts of turbidity in water.

Temperature. The temperature of the water supply as it comes to the ion exchange system and the temperature variations with seasons of the year are important and should be noted with the water analyses and recorded regularly. This temperature measurement is of importance to ion exchange not only because of its effect on the speed of reaction or exchange (particularly with the weakly acidic exchange resins and the weakly basic anion exchange resins), but also because of the adjustment of backwash flow rates necessary to account for the viscosity change in water and the resulting changes in the resin bed expansion.

The effect on the exchange rate of even the strongly basic and acidic ion exchange resins at temperatures approaching the freezing point is significant, and adjustments in flow rates (or contact time) should be made to maintain capacity and quality.

Temperature effects on silica capacity and leakage of strongly basic anion resins are also significant. Contrary to most exchange reactions, which increase in reaction rate as the temperature goes up, silica removal and leakage are adversely affected by high temperature. First, the higher temperature tends to solubilize the silica not removed during the normal regeneration. Second, the presence of even small amounts of sodium leakage coming from the cation resin will result in production of sodium hydroxide in the anion resin, raising the pH and adversely affecting removal of the silica.

The effect on backwash bed expansion of ion exchange resins is that the lower the temperature, the higher the viscosity of water becomes and the greater will be the bed expansion with the flow rate remaining the same. The result will be the possible loss of resin from the tank, unless the flow rate is adjusted downward as the temperature decreases.

Conversely, with the higher temperature the water viscosity goes down, and if the flow rate remains unchanged the bed expansion will be less than that required to remove dirt and fine particles or to obtain good resin classification. Increasing the flow rate as the water temperature increases will remedy this problem. Resin manufacturers supply information on all resins, showing the relationship of water temperature to bed expansion at various flow rates.

Pretreatment of water. There are a number of pretreating processes used to reduce the suspended solids and TDS load to ion exchange systems. These include:

Chapter 3

- Filtration
- Coagulation and filtration
- Cold lime with or without soda ash
- Hot lime with or without soda ash
- Evaporator or distillation
- Electrodialysis
- Reverse osmosis
- Ultrafiltration
- Degasification
- Combinations of the above

The uniformity of water quality produced by some of the above processes is of course dependent on the uniformity of the water supply, but is also subject to variation in the operations of the processes and in the life of the component parts used in the membrane processes. These changes will also change the water analyses of the water being fed to the ion exchange systems that are used to obtain the quality of end product water desired.

Since some of these processes reduce the ionic load to the ion exchange systems, the run length or service cycle will be extended over that which would be obtained on the raw water with filtration. Because of this, changes in the ionic load by even relatively small amounts will show up in fairly high percentage amounts for the run length of the ion exchange systems.

For example, the amount of carbon dioxide following a degasification unit downstream of a reverse osmosis unit might be specified by design as 3 ppm, as CO_2. However, changes in operation may produce 7 ppm, while the remaining ionic load might remain constant at 20 ppm and result in a significant 17% reduction in run length.

Water analyses and monitoring of changes in the feed water to the ion exchange resin system are necessary to properly manage, regulate, and troubleshoot these combinations of systems.

Organics. The role of organics in water supplies and the methods for their analysis, removal, and effect on performance of the ion exchange resins has received considerable attention by way of articles, publications, and seminars, but still is one of the least precise problem areas in ion exchange water treatment. It used to be that well waters were not suspect of being a source of organics in feedwater to ion exchange systems, but with contamination of the groundwater becoming more widespread, it is advisable to include tests for organics in the water analysis.

A very broad range of substances in water have been placed under the heading of organics, including some which may have inorganic components as a part of the organic structures and some which are of borderline solubility or of colloidal nature. The individual identification of the exact nature of these organics and their concentrations in water are not easily accomplished. Even if a detailed analysis of these organics could be obtained on a water supply, the ability to rate the capacity for removal of them by various systems, including ion exchange, is not an exact science at the present state of the art.

Test methods employed for analysis of organics in water, such as COD (chemical oxygen demand), TOC (total organic carbon), gas chromatography coupled with mass spectrometry, and ultraviolet and infrared spectrophotometry, are useful in making one aware of the amounts and in some cases types of organics that may exist in a water supply so that precautions may be taken to employ the best means to reduce those problems.

Pretreatment with coagulation, filtration, chlorination, carbon, ultrafiltration, reverse osmosis, and ion exchange resin organic scavengers can be helpful in protecting the working ion exchange resin DI system from fouling, and in insuring its ability to produce high quality product water. The selection of ion exchange resins that are used in the operating DI system to best remove organics and resist fouling is also of importance.

Summary

It is not possible to overemphasize the importance of obtaining detailed, accurate, and frequent water analyses that take into account variations in water supplies to be treated by ion exchange resin systems. These analyses are important for the proper design of ion exchange resin systems as well as for their optimum operation.

Understanding how water analyses are expressed and converted to usable terms will be helpful in applying this information to calculations used to determine the capacity of ion exchange resins in water treatment and the quality of water that can be produced.

Reference Test Methods

For ASTM standard methods for analysis of water and testing of ion exchange resins, the reader is referred to:
AMERICAN SOCIETY FOR TESTING AND MATERIALS
1916 Race St.
Philadelphia, PA 19103

AWWA testing methods may be obtained from:
AMERICAN WATER WORKS ASSOCIATION
6666 W. Quincy Ave.
Denver, CO 80235

CHAPTER 4

ION EXCHANGE RESINS AND THEIR REACTIONS

Although ion exchange as a process was recognized in the mid-1800s, the first synthetic ion exchange resins that permitted deionization were not discovered until 1935. Complete deionization with the removal of weak anions like silica and carbon dioxide was not possible economically until the development of strongly basic anion exchange resins in the 1940s.

The further development of improved resins and processes to utilize them has proceeded very rapidly in this relatively short period of time. Now there are many millions of cubic feet of resin installed for water treatment and for other specialty applications such as the sugar, uranium, pharmaceutical, catalytic, hydrometallurgical, food, wine, and chemical processing industries, to name a few. Indeed, ion exchange processes have been responsible for the rapid growth of some industries, such as the electronics industry, which would not have been possible without ion exchange. However, like the electronics industry, much further growth can be expected for ion exchange resin applications, and there is need for greater knowledge by those working with them.

The chemistry of the ion exchange resins and their manufacture may seem complex to those without knowledge of the chemical terms that are used, but the following discussions can be helpful in giving an insight into the macroscopic and molecular worlds of the ion exchange resins. The history of ion exchange resins and the chemical composition of the earlier resins began with naturally occurring clays and zeolites. These zeolites or minerals had relatively low capacity and were practical only for softening water.

The Synthesis of Ion Exchange Resins

The next progression was to synthetic resin polymers utilizing the phenol formaldehyde resin structure as the insoluble base onto which the ion exchange site was attached. The phenol skeletal structure may be shown as:

```
        OH
        |
        C
      ⁄   ⫽
   HC      CH
   ‖       |
   HC      CH
      ⫽   ⁄
        C
        |
        H
```

C = Carbon atom
H = Hydrogen atom
O = Oxygen atom

The six-carbon-ring formation above is one that will be seen in many of the resin structures used in the preparation of ion exchange resins. The simplest structure as shown below is benzene and is called the benzene ring.

```
         H
         |
         C
      ⁄    ⫽
   HC       CH
   ‖        |
   HC       CH
      ⫽   ⁄
         C
         |
         H
```

For simplicity (without the hydrogen and unsaturated bonding) the above structures are shown as:

The above space-saving way of representing the benzene ring and phenol makes it easier to show how these organic molecules can be joined with a

Chapter 4

second ingredient to form a polymer such as phenol formaldehyde. The formaldehyde is shown as CH_2O. When reacted with phenol the units form a super molecule or polymer as shown below:

Monomers   Polymer

$$n \bigcirc_{OH} + n \; CH_2O \longrightarrow n \; H_2O + n \left[-\underset{OH}{\bigcirc}-CH_2-\underset{\underset{CH_2}{|}}{\underset{OH}{\bigcirc}}-CH_2-\underset{OH}{\bigcirc}-CH_2- \right]$$

n = number of molecules used and units in the polymer chain

A representation of a section of the polymer in a line drawing might look like this, a three-dimensional honeycomb structure:

Another way of visualizing this polymer formation from monomers is to consider the monomers as being links in a polymer chain, as illustrated below:

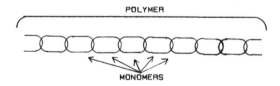

To obtain a larger structure and to give it strength to hold its shape, as well as to be insoluble, a number of these polymer chains are joined together by cross-links, similar to the way tire chains have cross-links which hold two lengths of chains together, as illustrated below:

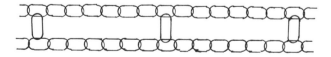

Differing from the tire chains, the polymers are cross-linked in a more random fashion, more like a box or ball of tire chains:

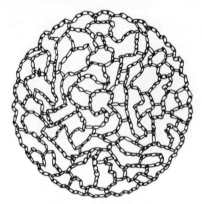

In a crude way, this might represent monomers polymerized to polymers and cross-linked into a copolymer resin bead. (A copolymer is a polymer formed from two different monomers such as styrene and divinylbenzene.)

In the case of styrene-divinylbenzene resins (the most common base for present-day resins) the monomers are:

The **C**=**C** represents what is called a double bond, which opens up like —**C**—**C**— with the two outside bonds ready to join with another monomer that has a double bond. This is done during the polymerizing process that requires heat and catalyst to form (in the case of styrene) the long-chain polymer called polystyrene (familiar to you in plastic toys and cups or tumblers), as illustrated below:

Chapter 4

In order to obtain a cross-linked polymer, divinylbenzene (DVB) is added to the styrene monomer and they are polymerized together. The divinylbenzene has reactive groups on both ends of the benzene ring and therefore can join to the benzene chain on both ends, linking two chains together as shown below:

```
     - C - C - C - C - C - C - C - C - C - C -
         ⏣       ⏣       ⏣     ⏣     ⏣   ⏣
       C - C - C      C - C - C       C - C - C
       |              |               |
     - C    C - C - C      C - C - C - C    C - C -
       ⏣    ⏣   ⏣     ⏣    ⏣   ⏣    ⏣
                       C - C - C
                       |       |
                     - C      C - C -
                       ⏣      ⏣
```

These cross-links give the resin structure its strength, insolubility, and resistance to melting or distorting over a wide temperature range, and they determine the tightness or porosity of the structure.

A fairly new resin structure (developed in the late 1950s) called macroreticular or macroporous (large pores) uses the same monomer ingredients to make the beads of resin, but also included is a third ingredient which is soluble in the monomers, but becomes insoluble in the polymer structure as it is formed. The third ingredient is then removed from the resin structure, leaving a resin bead that has both a continuous resin phase and a continuous pore phase, resulting in considerable internal surface area.

The previously described resin bead is called gellular because it looks, under very high magnification, like a slice of gelatin. The new resin structure differs by being very porous, with discrete physical voids, and under very high magnification appears like compacted microspheres of gel resin. See Figures 4-1 and 4-2. There are many more monomer and polymer structures used in the manufacture of ion exchange resins than those given here. They may be investigated in the reading references listed in the Appendix of this text.

Resin Bead Formation

Let us now look at the way in which this copolymer bead is formed into a nice spherical shape. You are familiar with the fact that oil and water do not mix into a solution, and that the oil will float on the water. If you try to mix them together by stirring vigorously, little balls of oil will be suspended in the water, but they will usually join together or coalesce, and separate into two layers as before if the stirring stops. If you add some soap or detergent to the water and again stir vigorously you can get a suspension of oil in the water that will be stable for some time. These droplets of oil in water are spherical.

The manufacture of the styrene-divinylbenzene ion exchange beads starts with the monomers of styrene and divinylbenzene, both of which, like oil, are insoluble in water. A small amount of detergent and some catalyst are added to speed the polymerization of the monomers into the polymer, and the ingredients are stirred into the water to get them suspended as small droplets in the water.

The speed of stirring, the shape of the vessel, the temperature, and several other considerations will determine the size of the droplets, and therefore the size of the copolymer beads that will be formed. For most ion exchange applications, a particle size in the range of 20 to 50 mesh (0.0331 to 0.0117 inches or 0.84 to 0.297 millimeters in diameter) is desirable. This size range has been chosen so as to:

- Obtain hydraulic characteristics that will give reasonably low pressure drop (pounds per square inch per foot of bed depth) in a packed bed, with the flow of water down through a classified bed (classified with the smallest beads on the top and graduating to the larger beads on the bottom).

- Have an acceptable amount of bed expansion when the water is run upflow through the bed for backwash (to classify the bed and remove fine particles or dirt from the resin bed).

- Maintain the largest surface area of the ion exchange beads and the best reaction rate at reasonably high flow rates per unit volume of resin.

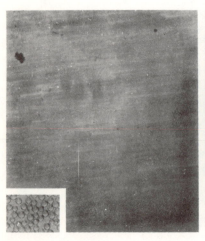
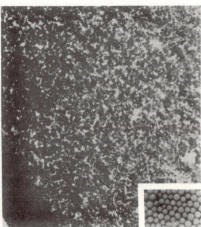

Figure 4-1. Electron micrograph of gellular (gel-type) ion exchange resin (50,000 X magnification).

Figure 4-2. Electron micrograph of macroreticular ion exchange resin (50,000 X magnification).

Photos courtesy Rohm and Haas Co.

Functional Groups or Exchange Sites

The polymer structures discussed to this point have to be further treated or reacted in order to add the ion exchange functional groups to the resin. There are some resins that are built with monomers which already contain ion exchange functional groups, but most of the commonly used resins have the functional or exchange groups added to the polymers or copolymer structures after they have been formed into the physical beads or granules.

Going back to the comparison of resin polymers to tire chains, we can now visualize the addition of ion exchange sites (or functional groups) to the resin polymer chains as the addition of spikes or small lumps on the individual chain links, throughout the large ball of cross-linked tire chains. These spikes or small lumps can be visualized as having a magnetic force built into them and being all of equal strength. In the cation resins the magnet will have the negative pole sticking out from the chain link to attract the positively charged cations, while the anion resins will have the positive charge sticking out from the chain link to attract the negatively charged anions, as pictured below:

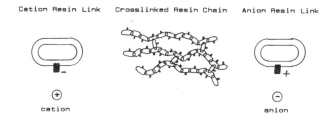

Putting the actual exchange functional groups on the polymer beads will depend on whether cation or anion resin is to be produced (or whether the exchange group was present on the monomer). In the case of cation resins, such as the strongly acidic resin, where the ion exchange functional group desired is the sulfonic group ($-SO_3^- H^+$), the copolymer bead is heated with concentrated sulfuric acid, which causes the sulfonate group to go onto the benzene ring of the polymer as follows:

Weakly acidic cation exchange resins can be prepared by polymerizing methacrylic or acrylic acid with divinylbenzene with the resulting copolymer having the desired functional group ($-COO^- H^+$) already in the structure.

For anion resins of the strongly basic resin class, where the quaternary ammonium ion exchange functional group ($-CH_2N(CH_3)_2{}^+Cl^-$) is desired, it is necessary to go through a two-stage process as follows:

Chloromethylation

$$-C-C-C- \; \bigcirc\bigcirc \; + \; 2\,ClCH_2OCH_3 \; \xrightarrow{catalyst} \; -C-C-C- \; \underset{CH_2Cl\;\;CH_2Cl}{\bigcirc\bigcirc} \; + \; 2\,CH_3OH$$

Amination

$$-C-C-C- \; \underset{CH_2Cl\;\;CH_2Cl}{\bigcirc\bigcirc} \; + \; 2\,(CH_3)_3N \; \longrightarrow \; -C-C-C- \; \bigcirc\bigcirc$$

with pendant $-CH_2-N^+(CH_3)_3\,Cl^-$ groups on each ring.

In the case of weakly basic anion resins or the type II strongly basic anion resins, other amines can be used following the chloromethylation step to achieve the desired anion resin.

The above brief discussion of the chemistry of synthesizing ion exchange resins is incomplete but is given as an illustration of the ways many ion exchange resins are manufactured.

For most water treatment applications, the cation exchange resins used are of the strongly acidic and the weakly acidic types. Strongly acidic cation resins can be compared in acid strength to sulfuric acid, while the weakly acidic cation resins could be compared to acetic acid (the acid present in vinegar).

The anion exchange resins used are strongly basic anion and weakly basic anion types. Strongly basic anion resins can be compared in base or caustic strength to sodium hydroxide (caustic), while the weakly basic anion resins could be compared to ammonium hydroxide (household ammonia).

The general distinction between strongly acidic cation (SAC) resins and weakly acidic (WAC) resins is that in the hydrogen form the SAC resins can split neutral salts, such as NaCl, by exchanging H^+ for the Na^+ of the salt producing HCl as the product, while WAC resins can not do so to any extent. This same comparative difference exists between strongly basic anion (SBA) resins and weakly basic anion (WBA) resins in the ability to split neutral salts by the SBA resin exchanging OH^- for Cl^- in NaCl to form NaOH, whereas the WBA resins can not do so.

Variations in the physical characteristics of the ion exchange resins to give differences in crush strength, degree of porosity, oxidation resistance, ability

Chapter 4

to resist organic fouling, particle size distribution, moisture content, swelling, hydraulic requirements, and many other considerations are all matters of concern in the synthesis and controls of manufacturing ion exchange resins.

Ion Exchange Resin Reactions

As has been noted in Chapter 2, the basic criterion to describe ion exchange reactions is that one ion to be exchanged had to be in solution and the other ion held on an insoluble solid (or in some cases in an immiscible liquid).

Greensands and sodium aluminum silicates (both natural and synthetic), which were the forerunners of modern ion exchange resins, were limited in their use to softening applications, as they were not sufficiently stable at low or high pH. The softening process is accomplished by starting with the resin or zeolites in the sodium form and then contacting the resin with water containing calcium and magnesium ions (hardness). The divalent calcium and magnesium ions exchange for the sodium ion on the solid resin, with the sodium ions going into the water solution in place of the calcium and magnesium, thereby removing the hardness and yielding soft or softened water.

By representing the ion exchange resin as R in the following illustration, the reaction can be represented as:

$$R\text{-}(Na^+)_4 + \begin{bmatrix} Ca^{++} \\ Mg^{++} \end{bmatrix} \begin{bmatrix} HCO_3^- \\ Cl^- \\ SO_4^{--} \end{bmatrix} \rightleftharpoons R\text{-}\begin{bmatrix} Ca^{++} \\ Mg^{++} \end{bmatrix} + (Na^+)_4 \begin{bmatrix} HCO_3^- \\ Cl^- \\ SO_4^{--} \end{bmatrix}$$

Notice that in the illustration a balanced equation is shown with four monovalent sodium ions on the resin being replaced with one each of divalent calcium and magnesium ions, and the resulting ions in the water will be balanced as to positive and negative ions as this is a necessary requirement in all ion exchange reactions. Obviously, the number of exchange sites on the resin will be more than those shown in the sample reaction equation, as this is only used to represent the reaction taking place.

Also note that the size of the arrows shows that the reaction is going predominantly from left to right, but there is a slight tendency for the reaction to reverse or to reach an equilibrium. For example, if the water and the resin were put in a jar or beaker and allowed to stand, the reaction would (over a period of time) come to an equilibrium where there would be some sodium ion on the resin and some calcium and magnesium in the water.

If the water were poured over a column of resin (held in a tube or column by porous plates or screen) and the water removed from the bottom of the column, then the reaction would be driven more to the right, since the reaction products (on the right) would be removed as they are formed and there would be little driving force to push the equilibrium to the left. Also, as the water is passing down the column of resin, there is more sodium on the resin toward the bottom and the water has less and less calcium and magnesium ions left in it, so the driving force would shift the reaction from left to right to an even greater extent or degree of completion.

The first condition described above with the resin and the water sitting in a jar is called *static*, since the water and the resin are not moving in relation to each other. The second condition described with the resin in a column and the water running through the resin is called *dynamic*. Most commercial ion exchange operations are done under dynamic conditions.

There are a number of other conditions that affect ion exchange reactions and influence the direction in which the reaction will proceed and the degree of completion obtained. Some of these are:

Temperature. Generally, the higher the temperature the faster the reaction proceeds. However, temperature has an influence on the solubility of some substances and, depending on the direction you wish the reaction to go, may have an undesirable effect.

Time. Under static conditions the exchange reaction proceeds rapidly at first and then slows down as the number of exchangeable ions on the resin decreases, until a true equilibrium point (when the reaction stops), which is dependent on the particular ions (on the resin and in solution) involved.

Under dynamic conditions the exchange is driven to completion as the water moves down the column, but the degree of exchange is also dependent on the time of contact of the water solution with the resin in the column (or gallons per minute (gpm) per cubic foot (CF) of resin).

Depending on the ions involved, there is an upper limit to the speed of flow per unit volume of resin in order to get a reasonable or acceptable degree of exchange to occur. There is also a lower limit that would be acceptable, since at very low contact time or flow per unit of resin volume, the dynamic conditions would be approaching those of static conditions.

Concentration. In the above illustration of an ion exchange reaction, there is no indication of the amount or concentration of each substance in the reaction, and this is the most important information needed to determine if the reaction will take place, and the direction and extent the reaction will go. In other words, the size of the arrows from the left to right or right to left is determined by the concentration of each ingredient in the reaction.

A large majority of the waters to be treated by ion exchange are in the range of 50- to 500-ppm concentration of TDS. For purposes of this discussion, let us use 200-ppm TDS and a water with a hardness of 171 ppm as $CaCO_3$, or:

$$171 \div 17.1 = 10 \text{ grains/gal as } CaCO_3$$

to be softened by a cation resin with a capacity of 30 Kilograins (Kgr) as

$CaCO_3$ per CF of resin. Since 1 CF of resin is equal to 7.5 gallons and the void volume (the space not occupied by resin in a packed column of resin) of resin is approximately 40%, the total water in contact with that CF of resin at any time would be:

$$7.5 \times 0.4 = 3.0 \text{ gal.}$$

The amount or concentration of hardness in the water exposed to 1 CF of resin would be:

$$3 \times 10 = 30 \text{ grains as } CaCO_3.$$

The CF of ion exchange resin with a given capacity of 30 Kgr/CF would have a concentration of 30,000 grains of sodium ions to exchange for the 30 grains of hardness in water, or a ratio of 1,000 to 1.

It would not be hard to see that the driving force to remove almost all of the hardness from the water and exchange for the sodium on the resin would be present and would move the reaction from left to right very strongly. This would occur because of the much stronger affinity of calcium than sodium for the resin exchange sites and the larger number of available exchange sites on the resin.

Now consider a situation where the same resin that removed hardness from the water (in the exchange for sodium on the resin) was mostly in the calcium and magnesium form. Suppose it were to be regenerated back into the sodium form, so that it could be reused in the softening process. If we follow the same logic of increasing the concentration of sodium ions on the right side of the illustrated reaction to shift it to the left or to put the resin back in the sodium form, we find it can be done, as illustrated below:

$$R-(Na^+)_4 + \begin{bmatrix} Ca^{++} \\ Mg^{++} \end{bmatrix}(Cl^-)_4 \rightleftarrows R-\begin{bmatrix} Ca^{++} \\ Mg^{++} \end{bmatrix} + (NaCl)_4^*$$

* Shown is a balanced equation, but in regeneration high concentration and quantities of salt are used.

To get the reaction to move right to left we use a sodium chloride brine solution at about 10% concentration to do this. One void volume of 10% brine (85,000 ppm $CaCO_3$) is equivalent to 14,912 grains of sodium as $CaCO_3$. In addition, the total amount of sodium chloride used (at 15 pounds per CF) would be about three times the amount of hardness on the resin. As can be seen, the driving force from the preponderance of $Na^+\ Cl^-$ to reverse the reaction is considerably greater than that which was present in the softening process, so the regeneration can be accomplished.

It is important to use dynamic flow over the resin to sweep away the calcium and magnesium removed or regenerated from the resin so as to reduce those ions from the other or left side of the reaction, so that their effective concentration is reduced as the regeneration proceeds.

The above discussion of the reactions of ion exchange resins is given as an example of the exhaustion, loading or service cycle, and regeneration cycle of both cation and anion exchange resins with other ions. There are differences in the affinity of the other ions for the resins or resins for the ions, which are touched on in Chapters 7 and 8.

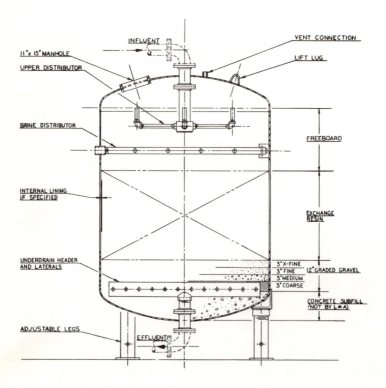

Figure 4-3. A sectional elevation drawing of a typical 72" diameter softener.
Drawing courtesy L*A Water Treatment Corporation.

CHAPTER 5

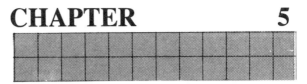

EQUIPMENT FOR ION EXCHANGE

Now that some of the basics of ion exchange resins and water have been discussed, an equally important ingredient in the ion exchange water treatment must be considered: the equipment that contains the resins and in which the ion exchange process can be performed.

The companies who manufacture the equipment in which the ion exchange resins are used must concern themselves with the proper engineering design of that equipment. Proper design will allow the ion exchange resins to be fully utilized for the job they were designed to do. Of concern are not only the tanks to contain the resins, but more importantly the internal distribution systems, the measuring systems to control the proper level and concentrations of the regenerating chemicals, the instrumentation required to monitor quality of the treated water, and proper hydraulic conditions. Also important are reliable valves and controls, sizing of piping, and proper installation of the equipment and ion exchange resins. These are only a few of the many requirements for properly functioning ion exchange systems.

Ion Exchange Resin Tanks or Vessels

Most all industrial ion exchange systems use vertical cylindrical pressure vessels of steel construction. There are a few (and some very large) gravity softeners constructed of concrete in rectangular shape, but flow rate and space constrictions have limited the use of this type of construction. There are also conical, spherical, and compartmented vessels used in special circumstances.

For softener tanks, where only salt brine is used for regeneration, a galvanized, Bitumastic®, phenolic, or epoxy interior coating can be used. For other ion exchange tanks where acids and caustic are used for regeneration, interior rubber lining is usually used. In special cases Saran® lining, polyvinyl chloride (PVC) lining, stainless steel, fiberglass, or steel tanks lined with food-grade rubber are used. The quality of the rubber and the effectiveness of the bond to the steel tank are of great importance to the usable life of the coated tank. The continuity of the coating to all openings and fittings in the tank, so that no voids are present, is a must, or concentrated corrosion can take place. A spark test is usually required to insure continuity of the lining.

Tank design pressures of 50, 75, 100, and 125 psig (pounds per square inch gauge) ASME code or noncode may be specified, depending on local pressure requirements. All ASME vessels are hydrostatically tested by 50% over design pressure.

In considering the design of the tank or vessel it is necessary to carefully review all the required openings or connections that may be needed, such as manholes (for easy access to install or repair internals), flanged inlet and outlet connections, regenerant inlet, resin removal connections, vent connections, instrumentation ports, sight glass ports (which should have sight glasses mounted flush with the interior wall), necessary distributor mounting brackets, etc.

Sizing of Resin Tanks

Sizing of the tank or vessel is dependent on the flow rates required, the amount of and types of resins to be used, differences in the freeboard (space necessary for backwash above the settled surface of the resin bed and top distributor), and allowances for expansion or shrinkage of the resin in its cycling. Other considerations are the capacity of the resin and therefore volume of resin required for the service run length desired, and the physical limitations of the building or space where tanks are to be installed.

Size of the tanks can range from diameters of generally not less than 30 inches (for industrial applications) to 12 feet. The larger diameters are generally restricted by the feasibility of transporting or shipping the completed tanks to the job site. Larger tanks can, however, be constructed on the job sites. An 8-foot-diameter tank is the largest generally used where highway transportation is required or for readily skid-mounted packaged installations. The size of readily available dished heads for the vessels to be designed is also a practical limitation.

Physical and Hydraulic Constraints

The height of the tanks or, as generally stated, of the straight sidewall, is limited to the 30-inch minimum bed depth recommended for ion exchange resins (18 to 24 inches of resin bed depth have been used, but are not generally recommended); plus the depth of any underbedding (graded gravel, anthracite, silica sand, or other coarse media of high density), which can range from a minimum of 6 to as much as 16 inches. With screened distribution laterals, which are becoming more widely used, the space occupied by the underbedding can now be used for added resin capacity. The capacity of the resin around the bottom distributors may not be fully utilized and should be discounted to some degree in figuring the volume of resin required. From the above considerations and allowing for a minimum freeboard of 40%, it can be seen that the minimum straight sidewall would have to be approximately 48 inches.

Chapter 5

> 30" bed depth
> + 12" freeboard
> + 6" underbedding
> = 48"

The limit as to the maximum vessel straight side might be the physical constraints of the containment building, but most usually is the pressure drop from the inlet to the outlet of a single tank or vessel. With as many as four vessels in series in the total system, consideration must be given to total pressure drop.

As water flows through any pipe or vessel, there is a certain loss of pressure due to friction with the walls of the pipe or other restrictions to the flow, such as would be caused by inserting a pipe of smaller diameter between two sections of larger pipe. Likewise, resin beads packed in a tank cause loss of pressure (pressure drop) as the water flows down through the resin bed. The pressure drop also varies with the resin particle size and the particle size distribution; the degree of or lack of classification (from small particles on top to larger on the bottom, resulting from hydraulic classification during backwash); the amount of deformation of the resin beads due to the pressure drop across the bed (thereby reducing the void space between the beads); and, of course, the flow rate and temperature of the water. Pressure drop during the service cycle may also occur due to suspended solids or foulants in the raw water.

Where the resin beads interface or make contact with the distribution system in the vessel is also a point of significant pressure drop, since the resin beads can deform into the open spaces of the screen or slots of the distributor, causing a reduction in the open area of the distributor.

There can also be considerable variation in pressure drop, depending on the type of ion exchange resins used. For example, a range of 0.7 to 1.0 psig in pressure drop per foot of bed depth can be found for most commonly used resins at a flow of 8 gpm/ft^2 and a temperature of 60°F. There are a few resins with pressure drop as high as 2.0 psig pressure drop per foot of bed depth and some with as low as 0.6 psig. The pressure drop due to the distributor systems (including the resin interface with the distributor), and depending on the distributor design, can be in the range of 2 to 4 psig.

The generally accepted maximum of 10 psig pressure drop across the resin bed itself plus the distributor pressure drop of 2 to 4 psig gives a total value from the inlet to the outlet of 12 to 14 psig, which is satisfactory. This would mean that the maximum resin bed depth could range from 10 to 14 feet at a flow rate of 8 gpm/ft^2 at 60°F. Generally, however, because of height restrictions in the containment building and to give some allowance for resin fines, temperature change, and filtration of particulates, the normal bed depths used are in the range of 4 to 9 feet. Since the required room for backwash bed expansion is from 40% to 100% of bed depth, it can be seen that the straight wall can vary from 5.5 feet to as much as 19 feet.

Flow and Capacity Requirements

Another consideration in sizing of the ion exchange resin tank is the flow of water to be handled, along with the amount of resin needed to obtain the desired capacity or run length. As an example, if the water being treated has a very low ion load to be exchanged and the flow rate to be treated is very high, then the volume of resin required to provide sufficient capacity for a reasonable run length would not be large enough to come within the reasonable range of either superficial linear velocity (gpm/ft^2 of tank area) or possibly space velocity (gpm/CF of resin volume) to be satisfactory. Superficial linear velocities (gpm/ft^2) in the range of 4 to 10 gpm/ft^2 are considered reasonable, with 8 gpm/ft^2 normally used for most design work. Space velocities in the range of 0.25 to 5 gpm/CF are considered reasonable, with 2 gpm/CF used for most design work.

Here is another example. Where the ion exchange load of the water to be treated is high in comparison to the above example, while the the flow rate is the same, then the volume or cubic feet of ion exchange resin must be increased in order to provide sufficient capacity to give a reasonable run length. To increase the bed depth while maintaining the same diameter tank as might be used in the first example might pose a problem of too high a pressure drop. The solution to this problem might be to increase the diameter of the tank to take care of the larger volume of ion exchange resin, or possibly to go to two smaller tanks operating in parallel to give the desired square feet of tank area.

To illustrate the above discussions, let us take the first example, where the ionic loading is low and the flow rate high. Consider a water to be softened that has a low hardness of 50 ppm as $CaCO_3$ (or 2.92 grains per gallon). The flow rate desired is 100 gpm, and the capacity of the cation resin used is assumed to be 20 Kilograins per CF.

The 2.92 grains per gallon (gr/gal) $\times$ 100 gpm = 292 gr/min. In one day the hardness capacity required would be:

292 gr/gal $\times$ 60 min $\times$ 24 hr/day = 420,480 gr/day, or 420.5 Kgr/day.

To provide this capacity with a cation resin with 20 Kgr/CF capacity, 420.5 Kgr/day $\div$ 20 Kgr/CF = 21.0 CF of cation resin would be required for one day's service. However, at 100 gpm flow rate, the space velocity would be 100 gpm/21.0 CF = 4.8 gpm/CF, slightly on the high side, but acceptable. The linear velocity would be calculated by first assuming a minimum bed depth of 30 inches or 2.5 feet, which would give us 21.0 CF/2.5 ft = 8.4 ft^2 bed area. Then, 100 gpm/8.4 ft^2 = 11.9 gpm/ft^2 of bed area, on the high side of linear velocity. Using the design value for linear velocity of 8 gpm/ft^2, the

100 gpm/8 gpm per ft^2 = 12.5 ft^2 would be the square-foot area desired for the tank. A 4-foot-diameter tank would provide a surface area of 12.5 ft^2 and with a minimum bed depth of 2.5 feet would require 12.5 ft^2 × 2.5 ft = 31.25 CF of cation resin. This size of tank and CF of resin would then give a space velocity of 100 gpm/31.25 CF = 3.2 gpm/CF, in good range for this criteria, and a linear velocity of 8 gpm/ft^2, so all hydraulic conditions would be met.

To complete this illustration of the sizing of a tank to meet hydraulic and capacity requirements, allowances must be made for the backwash expansion of the resin, which in this case should be a minimum of 40%. To be conservative we will round off the volume of resin from 31.25 to 32 CF. Now the bed depth will be 32 CF/12.5 ft^2, with 40% freeboard required for backwash bed expansion, 2.5 ft × 1.4 = 3.6 ft, or rounded off to a 4-foot straight side wall would be adequate. This 32-CF unit would be capable of treating 1.52 days of continuous water requirements at 100 gpm before requiring regeneration with sodium chloride brine.

Now let us compare the system design where the ionic load is high at a hardness of 500 ppm, in place of the 50 ppm used in the above example, and see how this would change the tank design and sizing, maintaining the flow rate at the same 100 gpm. Now the hardness load is:

500 ppm/17.1 = 29.2 gr/gal, which will be:

29.2 gr/gal × 100 gpm × 60 min/hr × 24 hr/day = 4,204,800 gr/day
or 4,204,800 gr/day/1,000 = 4,204.8 Kgr/day.

At a capacity of 20 Kgr/CF for the resin we would have a requirement for resin of 4,204.8 Kgr/day ÷ 20 Kgr/CF = 210.2 CF/day (This is not completely accurate. We will later learn how changes in ionic load or concentration of ions to be removed affects the capacity of the resin).

It can be seen that if we tried to use the same size tank as used in the first calculations with a 12.5-ft^2 diameter, the bed depth would have to be 210.2 CF/12.5 ft^2 = 16.8 feet, much beyond the normal bed depth. Going to a more reasonable bed depth, let us look at a larger 6.5-foot-diameter tank with an area of 33.1 ft^2. This tank would give us a bed depth of:
210.8 CF/33.1 ft^2 = 6.4 feet.
The superficial linear velocity would then be:
100 gpm/33.1 ft^2 = 3.0 gpm/ft^2 and the space velocity would then be:
100 gpm/210.2 CF = 0.48 gpm/CF, both criteria within reasonable values. The values could be rounded off to the conservative side with 211 CF of resin to give a bed depth of 6 feet, 4 inches. With 40% bed expansion for backwash required, that would give 6.4 ft × 1.4 = 9.0 ft straight side. This unit would produce water at 100 gpm for a 24-hour day.

There are conditions where much higher flow rates are used routinely, such as in condensate polishing. Condensate polishing is the term used for the removal of corrosion products (such as iron and copper oxides), as well as ionic impurities (that originate from cooling water condenser leaks) that contaminate the otherwise high quality condensate. Ion exchange resin beds can be used both to act as a filter to remove the particulate material, and to deionize the ionic contamination. For low pressure boilers, sodium-form cation exchange resins alone can be used satisfactorily to treat the condensate return, since sodium and salt leakage are not that critical, but in ultra high pressure systems, both cation and anion resins are used (generally in mixed-bed systems) to filter and deionize the condensate.

Special resin particle size cuts are used to reduce the pressure drop. The temperature of the condensate may be in the range of 120°F, which reduces the viscosity of the water and the pressure drop, so that flow rates of 50 gpm/ft^2 or higher can be used with resin bed depths of 3 to 4 feet, with good results.

There are several other criteria to be considered in the design of an operating system before coming up with the final tank design choices, and these are:

- Regeneration time, including backwash, regeneration injection, displacement rinse, fast rinse, and time for inadvertent delays.
- Maintenance or downtime for adjustment or repairs.
- Duplicate or multiple systems so that one unit or train is always available for service while another is regenerating.
- Variation in water analyses, which would affect run length.
- Variation in flow requirements.
- Storage tanks for handling variation in flow requirements, regeneration time, and water needed for regeneration purposes.

Distributors and Collectors

Several comments have already been made regarding the importance of distributors and their influence on the operation of an ion exchange system. Unfortunately, most discussions on ion exchange resin technology do not go into any great detail on the subject, but merely point to the importance of good distribution of flow to utilize the available capacity of the ion exchange resins installed in the tank. At this point the subject is then left up to the equipment manufacturers. Many of them have come up with sophisticated and patented designs, while some others have not.

Chapter 5   39

The requirements for good distributor design should include:

- Minimum pressure drop at the resin interface with the distributor.

- Sufficient open area in the distributor to handle the flow, while having sufficient back pressure so that the flow will be uniformly spread over the surface area of the vessel.

- Modification of (or additional) distributors to handle the regeneration slow flow rates as compared to the high flow rate of the service cycle and the normally high flows during backwash.

- Good support of distributor laterals to prevent distortion or breakage of laterals during changes in flows or, during backwash, possible movement of resins against buried laterals or against those just above the top of the resin bed.

- Proper support of screen-wrapped laterals and securing of the screen covering to the laterals to prevent rupture or ballooning when reverse flows are applied.

- Proper choice of screen or mesh weave to provide the maximum open space where resin makes contact with the screen and also to allow cleaning of resin from the screen on reverse flow.

- Proper placement of the holes in the laterals so that flow is directed to allow uniform coverage of the area of the tank. Placement of holes should also reduce dead spots where regenerants might accumulate and not be washed from the resin bed, and where water flow would not contact some of the resin during service flow.

- Resistance to chemical attack, corrosion, and high temperatures.

- In some cases, proper selection of screen size so that the resin fines or suspended solids can be removed during backwash without resin loss. (This is normally done to allow loss of finer than 50 mesh while preventing loss of 50 mesh or larger).

- If support media such as gravel, silica sand, anthrafil, or others are used, the size should be graded so that the smallest size media is next to the coarse resin on the bottom of the bed, so as to prevent resin from penetrating the support media, and so that the largest size media is covering the bottom distributor or collector.

Figure 5-1. Typical condensate polishers. Photo courtesy L*A Water Treatment Corporation.

The types of distributor or collector systems used fall generally into the following categories:

Header-lateral. The header is the large pipe through which the water, chemical, or air flows to the laterals, which branch off from the header at right angles on a horizontal plane and are of smaller pipe size. The laterals are spaced along the header and are of varying lengths to cover the area of the cylindrical vessel.

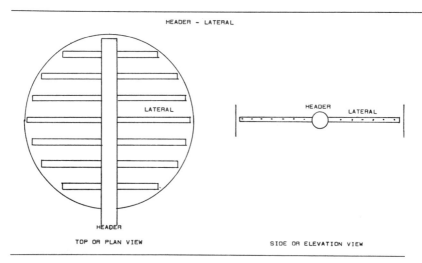

TOP OR PLAN VIEW   SIDE OR ELEVATION VIEW

Header-lateral distributors in the upper portion of the tank are supported by fittings and bracings (depending on the size of the tank) to prevent failures due to resin or water surges.

The types of screen and weave of screen, as well as how the screen is supported away from the surface of the pipe where the holes are drilled, all make important differences in pressure drop and distribution. Strainer heads or sections of well screen may also be used by being positioned along the laterals in place of screen wrapping.

Where the distributors are located in the top of the tank to distribute flow on the service run, the holes in the laterals may be positioned in an angle toward the top of the bed, vertically, or horizontally. The positioning of the holes must be designed so that there is no swirling or vortex action created by the incoming flow to disturb the top layer of the resin bed, as this would result in uneven flow through sections of the bed. Improper upper distributor design can be recognized by opening the top manhole after the end of the service cycle and before backwash, and inspecting the resin bed surface for hills and valleys.

Regenerant distributors, normally placed just above the top level of the resin bed in its most expanded or swollen condition, have a different requirement than the top distributor, since the chemical regenerant flows are most often at much lower flow rates than the service flow rates. Therefore pipe size, open area, screening, and placement of drilled holes must be modified to handle these different conditions. With the location of the regenerant distributor about 6 to 12 inches above the expanded or swollen resin bed, the distributor will be completely covered with resin during the backwash cycle and must be screened accordingly to prevent resin from getting into the regenerant distributor and causing some irregular plugging of the distributor.

The expansion of the resin bed during the start of the backwash (particularly if the backwash flow is started abruptly) may cause strains on the regenerant distributor: strains large enough to cause failure of the distributor if it is not well supported and braced.

In other resin systems, such as a mixed-bed system, the distributor may be buried in the resin bed. Here the strength of the laterals and their supports or bracing is of even greater concern, since the resin bed can act almost like a solid around the distributor. A sudden movement of the resin bed caused by changing the direction of flow upward or downward can cause a displacement or rupture of the laterals.

The bottom distributor may have the header-laterals positioned on a flat plate on the bottom of the vessel, where the head may be filled with concrete or other substances, and covered with a coating similar to that used for the tank. The steel plate can also be supported by bracing and can be welded to the side and bottom head of the tank.

The bottom or collector distributor design may be subject to an even greater number of variables, depending on the system design. In down-flow co-current operation, it is necessary that the bottom distributor must:

- Handle the backwash flow rate (high or medium).

- Handle the regenerant flow rate (low or variable).

- Handle the service flow rate (high).

- Be positioned so as to leave no dead spots or areas where there is little or relatively little fluid movement.

- In some cases, be able to handle air flows used to mix or scrub the resins.

Generally, the backwash and service flow rates may be within the range where the distributor may handle both satisfactorily, but the regenerant flow rate may be sufficiently lower so as to require a separate distributor system to maintain uniform flow over the entire area of the bed. However, if the resin bed has been properly backwashed to achieve good particle size classification of the resin and the regenerant distributor at the top is designed with good flow distribution, the bottom distributor used for service and backwash flow should then function satisfactorily. The bottom distributor or collector can be positioned on a flat false bottom (as discussed earlier) or with down-comers (small pipes coming down from the laterals as illustrated below) with nozzles or slotted circular strainers to prevent the ion exchange resin or resin support media from getting into the collector on the service flow and to improve the flow pattern on regeneration or backwash flow.

HEADER – LATERAL

WITH DOWN COMERS

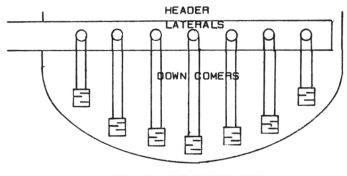

Hub and spoke. As the name indicates, this type of distributor looks like an old wagon wheel with the rim missing. The flow is directed to (or from) the hollow hub by a pipe through the side (or the bottom or top head) of the tank, and distributed out through the pipe spokes, branching out from the hub to near the walls of the tank. The mechanics of drilling the holes or screen coverage are handled in a similar fashion to that of the header-lateral distributors.

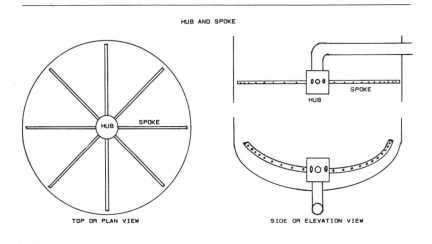

In the case of the bottom distributor with the hub and spoke type, the pipe spokes can be formed or shaped to the contour of the head to obtain good utilization of the resin in that section of the tank. Care must be taken with this arrangement, however, to be sure that the design takes into account the fact that the raised ends of the pipe spokes near the vertical wall of the tank will present a shorter flow path for the water or regenerant than the flow down the center of the tank. Also to be accounted for is that most of the resin volume is at the outer periphery of the tank, so that the open area of the pipe spokes must be adjusted accordingly to maintain uniform flow and the fullest utilization of the resin in the head.

Other distributor systems. In the above discussions we have covered two basic, popular distributor systems. In some cases they may be used together in the same tank. We have also made brief mention of the graded inert media underbedding. In this latter system the major problem that can occur is the upset of the graded media by air or gas liberation, so that the classification of particle size is disrupted and resin may be lost out of the unit. By controlling the flow for backwash by the outlet valve to keep some pressure on the system, the gases will tend to stay in the solution and will lessen the probability of this sort of disruption. Eliminating any fast-opening valves on the start of the backwash will also lessen the chances of disrupting the underbed media.

Strainers of various designs are also mounted on the heavily reinforced flat steel plates on the bottom of the tank. The strainers are spaced at even intervals on the surface of the plate to provide for good distribution, as illustrated in the accompanying figure.

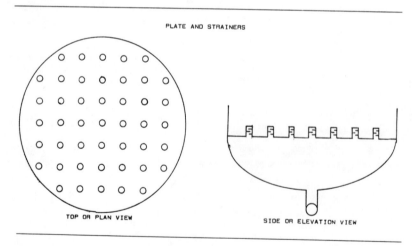

There are a number of other distributor and collector designs that have been used, but the above designs are probably the most popular and widely used.

Chapter 5

Face Piping and Valves

Controlling the direction and volume flow of water, and therefore of various regenerant and waste streams, requires some interesting and challenging engineering skills. This is particularly true where there are multiple vessels and trains of vessels with some units on line in service, while others are in various stages of regeneration.

The pipe and valve sizes will affect the pressure drop through the system (in addition to that in and out of the resin tank). The speed of opening and closing of valves is important to prevent sudden changes in flow, which could cause the resin bed to lift as a plug rather than slowly expanding to a loose fluid state on backwash. Also, rapid opening or closing of valves can create a hydraulic ram or water hammer, causing resin breakdown.

Valve choice is of great importance in order to insure positive closing, so that there will not be even small leakage of one stream to contaminate another in cross-pipe systems, such as a raw water line being isolated from the treated water stream.

In semiautomated or fully automated systems there are needs for fail-safe valves, which in the case of electrical, hydraulic, or pneumatic failure would prevent problems with intermixing of chemical regenerants, drainage of beds, or contamination of treated water. Positive block valves, properly placed, are a necessity to prevent contamination of other streams and back flows in regenerant systems.

Bulk Chemical Storage

There are six most generally used regeneration chemicals for ion exchange resins. The only one that does not pose a handling problem of any great degree is salt. In all other cases, the chemicals present some safety problems, and precautions should be observed because they are corrosive and dangerous to personnel and to the surrounding environment. Materials of construction of the bulk tanks, fittings, pipes, valves, pumps, measuring devices, and containment structures must be carefully engineered. Vents on the bulk tanks are necessary to prevent vacuum in the storage tanks as chemicals are being drawn out, and to prevent pressure when the tanks are being filled or when the fluctuation of temperature causes expansion and contraction. Some types of scrubber systems on the vent lines are necessary to prevent fumes from escaping to the atmosphere, and to prevent moisture and carbon dioxide from being brought into the tanks from the atmosphere. Handling of each of the more popular chemicals will be discussed in the following sections.

Salt (NaCl). In most cases the salt is delivered in bulk or bags. The salt is dissolved in water to make up saturated brine in brine tanks or large concrete basins. From there it is drawn or pumped to dilution or day tanks, or educted and diluted in line to the proper concentration for use in the regeneration process.

Sulfuric acid (H_2SO_4). 96% to 98% H_2SO_4 sulfuric acid in bulk is normally supplied at 66° Baumé (a hydrometer is used to measure density or specific gravity of solutions). In most large plants the concentrated acid is pumped to an automatic mixing station at a controlled rate and is there blended with a controlled amount of dilution water to get the proper strength or strengths for injection to the resin bed.

Hydrochloric acid (Muriatic acid, HCl). Hydrochloric acid is supplied normally at 20° Baumé (32% HCl) strength and can be fed directly with a block bleed system, as there is little heat of dilution from this concentration to the concentration normally used for regeneration. The major concern and problems occur with the materials of construction (stainless steel should not be used) for storage, piping, valves, and pumps, as well as with fumes from exposed tanks.

Caustic (Sodium hydroxide, NaOH). Caustic can be supplied in the flake solid form or more often as a 50% solution. The 50% solution is less troublesome to handle, since it can be readily pumped, but it must be kept above a temperature of 52°F or it will solidify. With solid flake caustic, the necessity for dissolving the solid in water with batch dissolving methods requires time-consuming manual control.

Soda ash (Sodium carbonate, Na_2CO_3). Soda ash is supplied in dry powder form and must be predissolved before being used, but it can be fed by an automatic feeder system into a dissolving tank with little difficulty.

Ammonia or Ammonium hydroxide (NH_3, Ammonia gas; or NH_4OH, Ammonium hydroxide). Ammonia is supplied as the 100% gas in pressure tanks or bottles and as the water solution, ammonium hydroxide (29.4% as NH_3 or 60.4% as NH_4OH). The gaseous ammonia does present some problems of fumes and cooling when released from pressure and dissolved in water, but the aqueous solution can be handled much in the same way as 50% sodium hydroxide is handled.

Degasification (Forced-Draft or Vacuum)

Degasifiers are used to remove unwanted dissolved gases from waters. There are two types normally used in the production of deionized water, with one of their main functions being to remove carbon dioxide and reduce the load of this anion from the water going to the anion exchange resin. Reducing this load to the exchange resin creates economic savings because less sodium hydroxide is required to regenerate the anion resin, and less resin is required.

As sodium hydroxide is the most expensive chemical used in the deionization process, any reduction in its use affects operating costs. However, there are some costs for operation of the degasifier:

Chapter 5

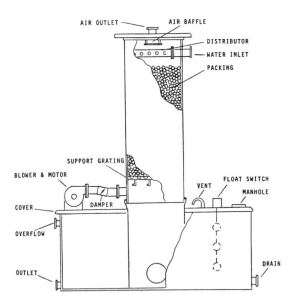

Figure 5-2. Diagram of forced draft decarbonator with catchment tank.

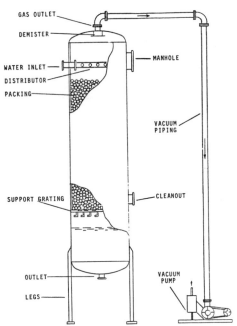

Figure 5-3. Diagram of single stage vacuum degasifier with vacuum pump.

- Equipment.
- Pumping. (The water pressure must be recovered by booster pumps to process the water through the anion exchange unit and on to service.)
- Blower or vacuum pump operation.
- Controls and maintenance.

As a general rule of thumb, if the alkalinity present in the feedwater is less than 100 ppm or 20% of the total anions, or the flow rate of water to be treated is less than 30 gpm, the justification for a degasifier might be questioned.

Strictly speaking, a forced-draft degasifier should be called a decarbonator, since its primary function is to remove carbon dioxide, while other gases such as oxygen are not removed, but actually increased to their saturation point. This increase in oxygen level has been well documented to cause shorter life expectancy for anion resins than when vacuum degasifiers are used. The latter device reduces the amount of all dissolved gases in the water.

Both forced-draft and vacuum degassing systems allow the water to fall through a tower filled with a series of trays or packing material. In the case of the forced-draft decarbonator, air is forced upward through the falling cascade of water to remove the carbon dioxide to around 10 ppm or somewhat less. In a vacuum degasifier, a vacuum created by a vacuum pump or a steam eductor is applied to the tower (with trays or packing) to remove carbon dioxide to around 5 ppm and oxygen to 0.2 ppm or less, as well as to reduce other gases to very low levels.

The forced-draft decarbonator takes air from the unit's environment, and even if it is equipped with air filters, it can be a source of contamination (collodial silica, solvents, bacteria, etc.) that will either foul the anion exchange resin or appear in the produced water.

Instrumentation

The number and types of instrumentation that are possible with ion exchange systems are dependent on the complexity of the system, as well as on the desired degree of automation and of control of quality and efficiency. As the complexity of the systems and the quality requirements become greater, the amount of instrumentation increases so that each unit in the system can be observed or recorded. This instrumentation is desirable in order to prevent malfunctions, to allow improvements in efficiency, to allow adjustments for changing feedwater quality, and to spot early warning signs of potential problems.

The field of instrumentation has gone through rapid improvements in quality and diversification, having paralleled technology developments of the semiconductor industry products. It would not be feasible to include detailed discussions of all the products and their application to ion exchange systems

Chapter 5   49

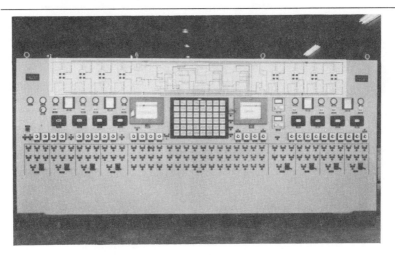

Figure 5-4. Photographs of typical instrumentation panels.
Photos ccourtesy L*A Water Treatment Corporation.

in this text. Listed here, however, are some of the areas where instrumentation is desirable, with the reasons why careful choice should be made to achieve the most reliable service.

Flow meters (or flow indicators). Flow meters are needed for:

- Keeping track of the total amount and instantaneous variation of water flowing through the ion exchange resin beds in the service cycle,
- Regulating flow during backwash cycle,
- Controlling flow of dilution water in regeneration,
- Controlling flow of regenerant chemicals,
- Controlling flow of displacement rinse water,
- Controlling flow of fast rinses,
- Measuring total waste volume, and
- Inventory purposes.

There are problems to be faced with the accuracy of flow meters or flow indicators, the first being temperature compensation. Water or water solutions, like all liquids, change viscosity (resistance of flow) inversely with temperature. Some devices to measure flow are not made to compensate for this change in temperature and therefore can not be accurate over the whole temperature range. When hot water or hot caustic solutions are used in regeneration cycles, for instance, there may be considerable errors not only in the flow meters or indicators, but also in the concentration measurements as recorded by hydrometers (density or specific gravity measurements). Without direct calibration (measuring volumes in containers versus time) or using temperature compensation devices, one cannot be sure of the actual flow.

Flow meters with movable mechanical parts, like most mechanical devices, can be subject to wear or fouling, and with age or exposure to fouling conditions they may become inaccurate. As with most instrumentation systems, periodic calibration checks should be considered.

Temperature indicators. Temperature of the raw water or process water is important for the reasons given above, but also is more critical where compensations must be made for flow rates used in the backwash cycle to assure proper bed expansion. With colder water the backwash flow rate must be lowered to assure that good resin will not be lost out of the top of the tank due to the higher viscosity of the water, or conversely, the flow rate should be increased when the water temperature is above the design level to assure that there is sufficient resin bed expansion to clean and classify the resin bed.

Temperature control is necessary for the water used to dilute caustic in the strongly basic anion regeneration step, where silica leakage is of importance. If temperature is too low, the silica leakage will be high and if the temperature is too high, it could cause resin damage and loss of capacity. It is therefore important to have temperature indicators both on the dilution water and on the diluted caustic supply line going to the anion resin, since dilution of caustic is exothermic (raises the temperature as the concentration supply is diluted with water) and it is important to keep the temperature of diluted caustic from going above 120°F. Short-time elevations to 130°F can be permitted with type I strongly basic anion (SBA) resins, but temperature limitations for type II and some acrylic SBA resins may be limited to 105°F.

Since 50% caustic is sometimes heated (to prevent solidification at temperatures below 52°F), a temperature indicator on the caustic supply line should also be used to be sure that the temperature is not above design. It is desirable, where hot caustic regenerations are used. to have a source of cold water tied into the hot dilution water supply, so that the temperature can be lowered if necessary.

When hot water is used to preheat the anion resin bed before hot caustic regeneration, a temperature indicator must also be located on the drain or waste line from the tank in order to see that the resin bed has been brought to the proper temperature.

Where weakly basic or acidic resins are used in an ion exchange system, it is also important to know the feedwater temperature, since these resins are more sensitive to temperature changes affecting the obtainable capacity. The feedwater temperature also has an effect on the silica leakage of strongly basic anion exchange resins. Higher leakage occurs with higher temperatures due to silica's greater solubility and the weak hold by the strongly basic anion exchange resin on the silica.

Another temperature-related factor is the problem of calcium sulfate ($CaSO_4$) precipitation during the sulfuric acid regeneration of cation resins. Temperatures over 85°F may cause more rapid crystallization (and precipitation) of calcium sulfate (gypsum) in the resin bed.

Conductivity meters. The conductivity of water is dependent on the ionic content of the water, specifically, on the ability of the ionic impurities in the water to conduct electricity. The conductivity of water is a direct linear function of the concentration of ionic impurities and therefore is a valuable tool in determining the amount of ionic impurities in the water, but it does not identify what they are. The conductivity measurement is dependent on temperature and pH. The higher the temperature the greater the conductivity, so temperature compensation is necessary.

The conductivity of neutral salts (such as NaCl) in water is less than the equivalent amount of either acids or bases (such as HCl or NaOH). It is therefore desirable to have a pH measurement in order to interpret the conductivity measurement in terms of concentration.

Conductivity can be expressed in terms of the reciprocal of resistivity (in ohms-cm) and called mhos-cm^{-1} or micromhos-cm^{-1} (1 micromho is one millionth of a mho). Put it another way, one micromho conductivity is equal to 1,000,000 ohms-cm of specific resistance.

A conductivity measurement on the incoming feedwater will catch variations in the ionic load that is to be removed by the ion exchange system, and therefore in the capacity in gallonage throughput that can be expected. One exception, however, is any variation in the amount of silica, which because of its very weak ionization is not shown in a conductivity measurement on the feedwater or on the treated water.

Conductivity measurements on the cation exchange resin effluent (when operated in the hydrogen form) will normally be higher than the feedwater conductivity, because the neutral salts in the feedwater are changed to acids:

$$R\text{—}H^+ + Na^+Cl^- \rightarrow R\text{—}Na^+ + H^+Cl^-$$

which are more conductive. In some cases a conductivity drop toward the end of the run has been used to signal the end of the service cycle. Measurements of a pH rise, however, might be more useful to signal the end point.

Conductivity measurements on the anion effluent are an excellent means of determining the end of the service cycle and monitoring the quality of water being produced during the run. With a weakly basic anion (WBA) resin system following a strongly acidic cation (SAC) resin system, the leakage of sodium from the cation exchange resin will show up as the neutral salt (NaCl) out of the WBA resin. If an SBA resin followed the cation exchange resin, the sodium leakage from the cation exchange resin would show up as sodium hydroxide (NaOH) out of the SBA resin with approximately twice the conductivity in μohms as would be the case with the WBA resin effluent. Furthermore, carbon dioxide (CO_2) would leak from the weakly basic anion exchange resin, but not from the strongly basic exchange anion resin, so that allowance must be made for the added conductance of this substance, in the case of the WBA resin. The signal of the end point for the service cycle would be a drop in pH and a conductivity increase for the WBA resin, and a drop in pH and a conductivity decrease for the SBA resin.

As silica is a very weakly ionized substance, it does not affect conductance to any great degree and therefore it can not be directly monitored by a conductivity measurement. Indirectly, however, there are some changes that do occur in the conductivity of effluent from a strongly basic anion resin that can signal the silica leakage break. Depending on a number of factors including co-current versus counter-current regeneration, organic foulants, degree of regeneration, and leakage of silica allowed, there will usually be a small but sharp increase in the conductivity (lower resistance), followed by a sharp decrease spike of conductivity (higher resistance) that rapidly returns to higher conductivity. As the anion resin bed's band of exhaustion nears the bottom of the bed, it brings about this blip in conductivity. Since the silica is only weakly held by the anion resin, it is pushed ahead of the more strongly held anions. The exhaustion bands are not distinct, however, and some of the other anions combine with the sodium leakage from the cation resin to form less basic (or more neutral) salts to show up as lower conductivity. There also can be some organic acids (if present in the raw water) that also may be weakly held and may be on the lead edge of the exhaustion band. This would account for the increase in conductivity.

To aid in anticipating the breakthrough end point of cation, anion, or even mixed-bed resin units, conductivity probes located in the resin tank approximately one fourth of the distance above the bottom distributor or collector in the resin bed are very helpful to prevent overrun of the service cycle. With the conductivity change that does occur (or possibly with a monitoring device for the specific ion that would be the first to leak at the end of the run) to signal the approach of the end of the service run, the operator will be alerted before the quality specifications are exceeded. This arrangement is particularly helpful in the case of silica leakage from the SBA resin, since conductivity break usually occurs after the silica break. Placement of the conductivity probe high enough above the bottom distributor (as determined by tests) will catch the silica leakage before it is above specifications. Conductivity instrumentation, used in this fashion, is quite generally more reliable than specific silica instrumentation.

Where very high quality water is required, mixed-bed ion exchange systems (a mixture of cation and anion resins) are used, and conductivity measurements in-line on the effluent stream are the best means of monitoring quality. Measurements must be in unexposed, flowing streams, as deionized water rapidly picks up carbon dioxide from the air and this will cause an increase in conductivity. The specific resistance of 18 megohms per cm^3 (or conductivity of 0.56 micromhos) represents exceedingly pure water and can be obtained with a carefully engineered mixed-bed system, but it is important that temperature compensators be a part of the equipment where accuracy is required, as the measurement is quite temperature sensitive.

pH meters. Measurement of the pH is a helpful tool to monitor performance of SAC resins in the hydrogen cycle (low pH) and SBA resins in the hydroxyl cycle (high pH) and to use in conjunction with the conductivity measurements. A pH rise can be used to signal an end of the service run on the cation resin, while a pH drop can be used on an anion resin.

pH measurements are also influenced by temperature and by the buffering effect of high TDS versus low TDS waters. Since pH measurements are a logarithmic function rather than a straight-line function, the measurement is more sensitive and therefore less precise, most especially in high quality water, as would be the case with mixed-bed systems where the specific resistance is above 1 megohm/cm^3 (or below 1 micromho conductivity). In-line probes for pH measurements are important since rapid pick up of carbon dioxide from the air will affect the pH, particularly in low TDS systems.

Level controls. Measurements of both the high and low levels of regenerants during chemical draw from concentrated or dilute tanks is important to insure that the proper amount of regenerant chemical is used in each regeneration cycle, as well as replaced for the next cycle. These controls can be either mechanical or conducting and can be tied into automated control of pumps.

Level controls are also required to maintain water level in the basins of degassing systems used for the removal of carbon dioxide or other gases by forced draft or vacuum.

There are some ion exchange systems where air domes (in place of water filling the freeboard space above the resin bed) are used, and in these cases the level controls are used to maintain a given depth of water over the top of the resin bed.

Level controls are also used to monitor water level in the product water storage tanks that are used to maintain continuous feed to plant services while an ion exchange unit is off line for regeneration, to handle irregular or maximum flow requirements of a plant, or to start up ion exchange systems in order to refill the storage tank.

Interlock systems. Where there is more than one ion exchange system or train, the interlock system prevents a unit or train from going into the service cycle before the other unit or system has completed its regeneration cycle and has been rinsed to quality (usually controlled by a conductivity signal).

Specific ion monitoring. As stated earlier, the number and types of instrumentation have rapidly progressed to a point where it is possible to monitor or record the amount of almost any ions that would be of interest in controlling leakage or run length, to a specific end point. The degree of reliability is improving, but a frequent schedule of inspection and calibration is still usually required.

Continuous systems for the measurement of silica, sodium, hardness, chlorine, copper, iron, and many others are available, and some are accurate to the ppb levels.

Pressure gauges. Inlet and outlet pressure gauges on each ion exchange vessel provide information on pressure drop that may occur due to packed or fouled ion exchange resin beds or other restrictions that could reduce flow, put excessive pressure drop across the resin bed, or restrict outlet pressure of the system to the extent that distribution of the water to points of use may not be sufficient.

Timers and switches. Where automatic or semiautomatic operation is used, there are a variety of mechanically driven sequence timers and step switches used to control the various cycles in the backwash, chemical regeneration, displacement rinse, fast rinse, recycle rinse, and other steps that may be programmed. The controls are usually set up to allow regeneration of the systems in trains, but also may be controlled for each unit to be regenerated independently. The combination of systems that are possible is almost endless and therefore the automated control possibilities are equally complex.

Since there are numerous varieties and types of controls available and systems for manual, push-button-initiated, semiautomatic, and fully automatic operation, the choice of system and components is an area where close consulting with the engineering and manufacturing firms is necessary when choosing the ion exchange system best suited to the user's needs. It should be stressed that as the degree of automation increases in the ion exchange systems, there is a corresponding increase in the number of components used, thereby increasing the possibilities of some malfunctions or failure in each component. There are, in addition, a large number of variables within the ion exchange process itself, such as the water supply, temperature, degree of exhaustion, resin losses in capacity with age or physical losses, fluctuation in flow rates, etc., so that supervision and frequent inspection for possible maintenance requirements by a knowledgeable operator are mandatory for all ion exchange systems.

The "high tech age" has also introduced computers and/or solid-state devices to the control and program logic associated with ion exchange systems. These devices bring far better reliability and flexibility to the systems, but usually require a price premium over more conventional timer, relay, and logic components.

CHAPTER 6

ION EXCHANGE CYCLES

There are many modifications of the cycles used with ion exchange systems. The majority now in use are probably co-current systems (downflow service, downflow regeneration). To begin these discussions, let us follow the operations of an ion exchange system through the service run and then through the regeneration steps, including backwash, chemical regeneration, displacement rinse (or slow rinse), and fast rinse (or final rinse to quality).

The diagram shown in Figure 6-1 will be helpful in following the operation steps or cycles of the typical ion exchange resin system that will be discussed in this chapter.

Service Cycle

After a unit has been rinsed down to a water quality for the service cycle, there are several reasons why it may be shut down. If there are several units in the system, it may not be needed until later. Its effluent may be used as feed to complete the regeneration of another unit. (A cation unit's effluent may be used for dilution or rinse water for an anion unit). When a shut-down unit is started up again, it may be necessary to again go into the fast rinse cycle to drain, to insure that the required quality water is available. Before the flow is switched to service (or to the next unit), the rinse to drain must be continued (usually for a short time) until the quality end point is obtained.

A quality end point reached in an initial rinse may not be obtained immediately when flow is resumed after a shutdown because the initial dynamic flow conditions have been changed to a no-flow or static condition. The ions in the water have come to an equilibrium with the ions on the resin, and in the process, given sufficient time, have become somewhat more uniformly distributed over the entire resin bed. Some of the ions that need to be removed in order to meet a quality end point would be present on the bottom of the resin bed and would show up in the first quantity of water which is removed as the flow of water is again started. It should be remembered that this chain of events can happen whenever an ion exchange system is shut down while still in any part of a service run. The severity of the problem will depend on the

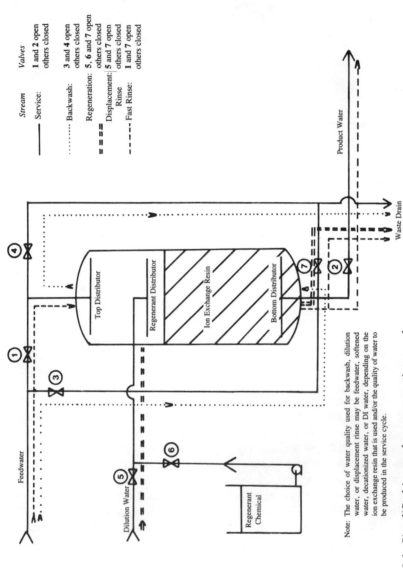

Figure 6-1. Simplified ion exchange operations cycle.

shutdown time, the degree of exhaustion, the particular ions involved, and the equilibrium that will result with the type of ion exchange resin being used. These precautions apply to both cation and anion resins.

With the ion exchange unit finally in the service cycle, let us look at what happens as we progress through the cycle, using as the first example one of the simpler ion exchange processes, that of water softening.

SAC resin (sodium form). Previously we reviewed the reactions of a strongly acidic cation exchange resin (SAC) in the sodium form as it exchanged sodium (Na^+) for the hardness in the water. Hardness in water consists mainly of calcium (Ca^{+2}) and magnesium (Mg^{+2}).

Let us look at the difference between calcium and magnesium in the selectivity of the resin for these two cations and consider that the SAC resin can be compared to sulfuric acid in acid strength and reactions. The calcium salt with sulfuric acid is calcium sulfate and the magnesium salt is magnesium sulfate. Looking at the solubility of calcium sulfate as compared to the solubility of magnesium sulfate, we find that calcium sulfate (gypsum) has a very low solubility, while magnesium sulfate (epsom salt) is more than 100 times more soluble. If we apply this comparison to the bond strength or selectivity of the Ca^{+2} ion for the SAC resin exchange site, we would expect that calcium would be held more strongly (be less likely to dissolve or hydrolyze off the resin) than magnesium.

As it turns out, this is indeed what happens, with calcium displacing magnesium as the service flow proceeds down the resin bed. With the cation resin in the sodium form, both calcium and magnesium (being divalent) displace the monovalent sodium from the resin and become attached to the exchange sites on the resin, while the sodium goes into the water and passes down and out of the resin bed. As more hard water is passed through the resin bed, the top resin layer would be converted to the calcium and magnesium form. However, since magnesium is not held as tightly to the resin exchange site as calcium, the calcium would displace the magnesium farther down the resin bed so that the very top layer of resin would become predominantly calcium form, with the magnesium in a layer below that, and the remainder of the column or resin bed would be in the sodium form as shown in Figure 6-2.

As the service run continues, the calcium band or layer would continue to expand down the column and displace the magnesium band, which in turn continues to expand farther down the column of resin and displaces the sodium. These bands of calcium and magnesium are not completely distinct, since that will depend on flow rate and uniformity of flow distribution. The higher the flow rate, the less defined or distinct the bands will be, and the more they will be spread out down the resin column. The flow of water down the column of resin will take the path of least resistance, so that there may be

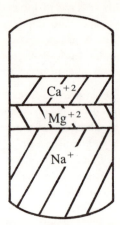

Figure 6-2. Ion exchange in softening resin column.

portions of the resin bed which will have become compacted (possibly due to improper backwash). Thus the bands or layers of exhausted resin may tilt or bulge. If this is the case, then as the end of the service cycle approaches and depending on the end point for hardness leakage (the amount of an ion or ions coming from a resin bed during the service cycle) and on whether there are some irregularities in the exhausted band, there will likely be an earlier breakthrough or leakage than if the exhausted bands are horizontal and flat.

Of course, the higher the hardness leakage end point permitted, the greater the length of the service cycle and the utilization of the capacity. The breakthrough should be fairly rapid with a well-designed system that maintains uniform distribution.

SAC resin (hydrogen form). The only differences in hydrogen-form operation versus sodium-form operation in the service cycle with SAC resins is that in the hydrogen cycle, all cations in the water (which will now include sodium in addition to the calcium and magnesium) must be exchanged for the hydrogen (H^+) ion on the resin. Since the sodium is monovalent and not as tightly held on the resin as calcium and magnesium, we will have a band of sodium below the magnesium band. It will spread out even more than the calcium and magnesium bands, and the remainder of the bed will be in the hydrogen form as illustrated in Figure 6-3.

The leakage of sodium would therefore occur first and would be the control end point of the service cycle. The sodium leakage is also influenced by both the TDS and the alkalinity of the feedwater as well as by the percentage of total cations that are sodium.

The TDS of the feedwater (or more precisely the total cations) will make a

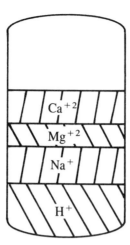

Figure 6-3. Ion exchange in decationizing resin column.

difference in the amount of hydrogen ions (H^+) that are exchanged for the cations being removed. These cations then combine with the anions present in the water to form the corresponding acids, which will increase in concentration as the TDS increases.

Even though the concentration of these mineral acids is relatively low, the sodium's weak bond to resin (as compared to calcium and magnesium) will be further weakened by the presence of these acids. Then some sodium will move on down the resin bed, spreading out the sodium band, and earlier and higher sodium leakage will occur.

The lower the mineral acidity (Cl^- and SO_4^{-2}) of the feedwater and the higher the alkalinity (primarily HCO_3), the lesser will be the amounts of strong acids formed and the more weak carbonic acid (H_2CO_3) will form. Since carbonic acid is such a weak acid that it will not cause sodium to be hydrolyzed or exchanged off the resin as readily, the sodium leakage will decrease and the achievable capacity will increase.

The higher the ratio of sodium to total cations, the larger will be the spread of the sodium band below the calcium and the magnesium bands, resulting in earlier breakthrough of sodium and greater sodium leakage throughout the service cycle.

The flow rate used during the service cycle does not have a great effect on the leakage values over the normal design ranges of 0.25 to 5.0 gpm/CF (normal design at 2 gpm/CF) with SAC resins, but there is some increase as flow rates go higher. As the sodium percentage of total cations increases, there will be a greater effect on the leakage from increased flow rates.

Temperature of the feedwater during the service cycle does not greatly influence leakage or capacity with the SAC resins, except at extremely low temperatures near freezing, where the reaction rates of the ion exchange slow down to reduce the run length and capacity by influencing leakage of the sodium.

WAC resin (weakly acidic). Thus far, we have been discussing the service cycle of SAC resins only, but there are increasing amounts of WAC resins being used in water treatment and there are some distinct differences in their performances that need to be understood. For now, we will consider those differences as applied to the service cycle only and cover the differences in other cycles as we come to those sections.

First, the WAC resins have a different exchange or functional group attached to the polymer backbone or structure than do SAC resins and this is a carboxylic group, chemically represented as RCOOH, which is of the approximate acid strength of vinegar (CH_3COOH). The hydrogen ion that is available for exchange is shown on the carboxylic group as $RCOO^-H^+$.

The selectivity of the SAC resins for the common cations can be listed in order of decreasing affinity or selectivity for the resin and in dilute solutions, with calcium as the most tightly held, magnesium next, sodium less, and hydrogen least. Descending order of affinity for WAC resins would have hydrogen most tightly held, calcium next, magnesium less, and sodium least. The main difference is that hydrogen goes from being the least strongly held on the SAC resins to being the most strongly held on the WAC resins.

This attraction (or affinity) of the carboxylic functional group on the WAC resin for hydrogen is important to remember when looking at uses for these resins. If the hydrogen ion concentration is high in the water solution (the pH is low) there is little reason for the carboxylic group to give up its hydrogen ion in an ion exchange reaction, but if the hydrogen ion concentration in solution is low (the pH is high) the carboxylic group will readily give up its hydrogen ion for the other cations in solution.

In a neutral water containing sodium chloride, if an exchange were to take place with a WAC resin, the reaction would be represented as:

$$RCOO^-H^+ + Na^+Cl^- \rightleftharpoons RCOO^-Na^+ + H^+Cl^-$$

With the formation of hydrochloric acid (H^+Cl^-), a strong acid, the pH would drop and the WAC resin would rather be in the hydrogen form, and therefore the reaction would not move to the right to any great extent, as the relative size arrows would show.

Another example is of a water containing sodium bicarbonate. The pH would be above 7, and the reaction with a WAC resin would be:

$$RCOO^-H^+ + Na^+HCO_3^- \rightleftharpoons RCOO^-Na^+ + H^+HCO_3^-$$

Here the product of reaction is carbonic acid, a weak acid, which breaks down to water and carbon dioxide, as follows:

$$H_2CO_3 \rightarrow H_2O + CO_2.$$

By degassing, the carbon dioxide is driven off, resulting in a near-neutral pH. Any other cations associated with bicarbonate in the water could be removed for the same reasons, by a WAC resin with a good capacity.

When the WAC resin is in the sodium form as shown above, the reaction of the WAC resin with sodium bicarbonate could cause a further reaction to take place by removing any calcium or magnesium and thereby softening the water. This reaction is similar to the softening reaction which occurs when the SAC resin is in the sodium form, exchanging the sodium for the calcium and magnesium. However, in the case of the WAC resins, the affinity of the carboxylic exchange groups for divalent cations is considerably greater than the sulfonic exchange groups on the SAC resin, and there will be less leakage of hardness.

In fact, sodium chloride brines can be softened with carboxylic resins, while sulfonic resins can do only a partial job. An explanation of this might be found by comparing the acid counterpart of the SAC resin to sulfuric acid and the WAC resin to carbonic acid. The salt formed by calcium with the sulfate anion is calcium sulfate (very insoluble), while the salt formed by calcium with the carbonate anion is calcium carbonate, an even more insoluble salt by a factor of over 100 times that of calcium sulfate.

One other difference that needs to be considered between the WAC resins and the SAC resins is the considerable change in volume (or swelling and shrinkage) that occurs with the WAC resins when they go from one ionic form to another. For example, SAC resins have a range of 5% to 8% change in volume in going from one ionic form to another, while WAC resins may change from 20% to 100% in volume. This difference can require changes in equipment design to accommodate the larger changes in volume of the WAC resins, both in the service cycle (swelling) and in the regeneration cycles (shrinking). This swelling in the service cycle can also cause a potential problem in pressure drop across the resin bed, particularly at high flow rates, so it is best to keep the bed depth low as compared to the diameter.

There are a fairly large number of WAC resins on the market, varying in composition of the resin structure and in acid strength. These differences require careful examination before a choice is made for each desired application.

Going back to the service cycle operation of the WAC resins, we can see from the above discussions that the exhaustion bands of the calcium and magnesium should be tighter and the sodium less tight than those with the SAC resins. Generally, the applications for the WAC resins in water treatment will be more restricted to special water analysis situations.

Since WAC resins in the hydrogen form can only remove cations associated with alkalinity in water and will preferentially remove calcium before magnesium, and magnesium before sodium, (but will not have a great capacity to split neutral salts) it follows that waters with high alkalinity (generally above 20% of total anions) would be good candidates for application of these resins. This might be illustrated as shown by a line drawing where the length of the line represents the amount of each constituent present in the water, on an equivalent-weight basis, as follows with Water No. 1:

Water No. 1:

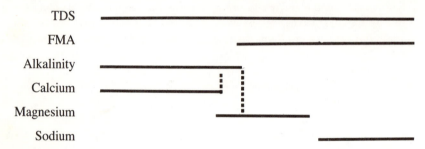

In the case with Water No. 1, all of the calcium and a small amount of magnesium but no sodium would be removed. Cation removal is limited to the alkalinity present.

Water No. 2:

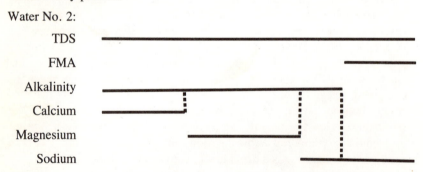

In the case of Water No. 2, all of the calcium, all of the magnesium, and about 40% of the sodium could be removed. The sodium, being less strongly held than either of the divalent cations, would be banded below the divalent cations and would be the first to show up as leakage. This leakage would show up as sodium alkalinity (or sodium bicarbonate, $NaHCO_3$).

If the purpose of using the WAC resin was to remove the cation load to a SAC resin that followed in a DI system, then the leakage of the sodium would not be of primary concern, as the SAC resin would readily remove that leakage at high capacity, so the service cycle of the WAC resin could be extended to a point where there would be magnesium (or hardness) leakage. Since the sodium ions on the WAC resin on or near the bottom of the resin bed would be displaced by magnesium and/or calcium in a softening process, the divalent ion load to the SAC resin that followed would be reduced or eliminated and would improve the capacity and regeneration efficiency of the SAC resin.

WAC resins have a high capacity for divalent ions and also have higher total capacity than the SAC resin, which can be utilized in the service cycle. We shall see later, in the regeneration section, that the WAC resins have much better regeneration efficiency than the SAC resins, which results in a more efficient system when the two resins can be used together.

Another use for the WAC resins is to soften waters which are not possible or practical to treat with SAC resins because of high TDS. To shift the equilibrium expressed by the following reaction:

$$R^{-2}\genfrac{}{}{0pt}{}{Na^+}{Na^+} + CaCl_2 \rightleftharpoons R^{-2}Ca^{+2} + 2NaCl$$

to the right, as in the softening process, there has to be a high affinity of the resin (in the sodium form) for the calcium, so the reaction will proceed from left to right. When the TDS in the water is very high (in the range of 10,000 ppm or 1%) then as the water is softened the amount of sodium chloride on the right side of the above reaction gets so high that it will try to force the reaction to go from right to left, (as would be the case with SAC resin) as if the resin system were being regenerated (often called back-regeneration). Because of the very large difference in affinity of the WAC resins for divalent ions over monovalent ions, WAC resins in the sodium form can soften even concentrated (but not saturated) brine solutions, where SAC resins can not.

The WAC resins also are more flow- and temperature-sensitive than the SAC resins, since the reaction rates or kinetics are usually poorer. Breakthrough of leakage can also be more gradual with WAC resins than with SAC resins, and monitoring the end point may be more difficult (depending on whether the hydrogen or sodium form of the resin is used and on whether hardness or alkalinity end point is desired).

SBA resin (strongly basic anion). The service cycle for the SBA resins is, to a degree, very similar to that of the SAC resins in that the major anions to be removed have different degrees of affinity or selectivity for resin being used and thus tend to band on the resin bed accordingly. Sulfate, being divalent, is held more strongly than chloride or bicarbonate, which are monovalent. Chloride is an anion of a strong acid (or mineral acid) while bicarbonate is an anion of a weak acid (carbonic acid $H^+HCO_3^-$), so that even though both are monovalent, there is a stronger attachment of chloride than of bicarbonate to the SBA resin. In each case the monovalent anions are less strongly held than are sulfate anions, and their bands extend to a greater extent down the column than the sulfate band.

Carbonate (CO_3^{-2}) and carbon dioxide (CO_2) may be considered as other forms of bicarbonate (HCO_3^-) or carbonic acid (H_2CO_3), as can be seen from the following:

$$CO_2 + H_2O \rightleftharpoons H_2CO_3 \rightleftharpoons H^+HCO_3^- \rightleftharpoons 2H^+CO_3^{-2}$$

Which forms are present will depend on the pH of the water, with low pH at the left side and progressing to the right side as the pH goes up.

For ion exchange applications (except anion dealkalization) the water is generally coming to the SBA resin from the SAC resin operating in the hydrogen form, (and perhaps even after having gone through a degassing device to remove carbon dioxide). The pH would be below 7, resulting in a majority of the bicarbonate form being present as the water goes to the SBA resin.

Silica (SiO_2), the most commonly occurring anion in water supplies, has been purposely left out of the discussion until now. This anion behaves considerably differently in how it loads or is exchanged onto the anion exchange resin (and regenerated off) than do most all other anions generally found in waters. Silica, being one of the major constituents (almost 28%) of the earth's crust, will be found in all waters, but because of its very low solubility it is generally present at under 30 ppm, with an average for the municipal water supplies of the United States at approximately 7 ppm.

Silica is considered to be in solution as the extremely weak silicic acid (H_2SiO_3 or H_3SiO_4), which is barely ionized and which is probably better represented as a nonionized species: $SiO_2 \cdot (H_2O)_n$ (n representing the unknown number for the water of hydration). As the amount of silica increases in the soluble form, there is a tendency for the silica to join together or polymerize into two or more units, and it can combine into a large number of units to become an insoluble polymer, which can be classified as nonreactive or colloidal silica. The polymerization of silica is enhanced at low pH. If nonreactive or colloidal silica is present in the water supply, it will not be removed to any great extent by the usual SBA resins. Special pretreatments required for its removal include chemical precipitation, ultra filtration, reverse osmosis membrane processes, and/or very-large-pore SBA resin.

Fortunately a majority of the silica in water supplies is of the soluble variety and can be removed by the SBA resins that are in the hydroxyl form R^+OH^-. The silica can be considered to be in the SiO_2 monovalent form with an equivalent weight of 60, as it is exchanged for the hydroxyl group on the SBA resin.

It must be understood that SBA resins do not have 100% of the exchange groups present as strongly basic groups, as with time, use, heat or oxidation attack, some of the strongly basic quaternary exchange sites may be changed to the weakly basic anion exchange sites of tertiary, secondary, or primary amines. These weakly basic exchange sites can remove the mineral acidity, but are not capable of removing the weakly acidic anion SiO_2^- from solution. The capacity for silica is therefore dependent on the amount of strongly basic (quaternary) exchange sites available on the resin.

Silica, being the weakest acid or anion usually present in the water, will of course be the most loosely held (by the SBA resin) of the normal anions, and will be in a broad band on the lower portion of the exhausted resin bed and be the first to show up as leakage from the bottom of the bed.

Factors influencing leakage of the silica during the service cycle are many, but probably the most important originate with the regeneration cycle and these will be discussed later in greater detail. Factors that originate in the service cycle include:

- Sodium leakage from the cation resin
- Service water temperature
- Ratio of silica to other anions
- Precipitation of silica in the cation resin
- Organic fouling of the SBA resin
- Flow rate

As was pointed out in the discussion of the service cycle of the cation resins, the leakage from the cation resins (in the hydrogen form) is sodium, as sodium chloride. When the sodium chloride goes through the SBA resin in the hydroxyl form, the chloride anion is exchanged for the hydroxide ion, producing sodium hydroxide. This sodium hydroxide, even in small quantities, raises the pH and has an effect on the loosely held silica in the lower portions of the anion resin bed. It causes some of the silica to be regenerated off the anion resin and appear as silica leakage in the effluent. The higher the sodium leakage from the cation resin, the higher the leakage of silica from the anion resin bed.

The solubility of silica is temperature influenced, in that the higher the temperature the more soluble the silica. Silica does tend to polymerize on the anion resin with time and as the pH drops in the exhausted bands where chlorides and sulfates are located. The dissolving of the silica polymer will occur as the temperature of the water goes up, resulting in a widening of the silica band to the point where some silica will leak from the resin bed.

As the ratio of silica to total anions increases (similar to the case for SAC resins when the sodium ratio to total cations goes up) the greater will be the spread of the silica band into the lower portion of the resin bed and the greater the chance for more and earlier leakage of silica.

Another less recognized cause of silica leakage, particularly where the silica ratio to total anions is high, is the possibility of precipitation of silica by polymerization of the silica in the low pH environment of the cation resin in the hydrogen form. This polymer form of silica in the cation resin can accumulate and continue to leak over to the anion resin and can be a source of higher-than-normal silica leakage unless it is removed by a cleanup with warm caustic wash (after a salt wash of the cation resin to remove any calcium or magnesium).

Organic fouling of SBA resins is a subject for more detailed discussions in Chapter 9 of this book, but in brief it is the retention of organic materials on the resin, which occupies or blocks the functional exchange groups and reduces the operational capacity of the resin. With fewer groups available to remove the silica, the silica leakage will go up. This problem is usually accompanied by longer-than-normal rinses to quality after regeneration and/or flow sensitivity with higher silica leakage at high flow rates.

The breakthrough or end point of a SBA resin service cycle is normally monitored by pH, conductivity, and/or a silica analysis. If there are mixed-bed polishers following the SBA unit, then pH and conductivity measurements are usually sufficient, since the mixed bed will handle removal of the low level leakage that might occur and not be seriously overloaded.

Thus far in the discussion of SBA resins we have not differentiated between the two types of SBA resins, type I and type II. Perhaps this would be the place to make this distinction, since in both the service cycle and the regeneration cycle there are differences that should be recognized. One of the differences is in the tertiary amine used to make the resulting strongly basic quaternary exchange group on the resin structure. The type I resins have better chemical and thermal stability, especially at high pH, and tend to produce higher quality water. The type II resins have slightly lower basicity and as a result are more efficiently regenerated with caustic than are type I resins. Because the type II resins have a greater percentage of weakly basic exchange sites than type I resins, they also have higher usable capacity for waters with high mineral acidity and low silica. Silica leakage of type II is generally higher than with type I resins. The type II resins, being slightly lower in basicity, are also more flow sensitive than type I resins.

WBA resins (weakly basic anion). The classification of WBA resins covers a wider variety of polymer structures and exchange functionality than most of the other resin groups. There are three types of functional groups that fall into the weakly basic category and they are (in order of increased basicity) primary, secondary, and tertiary amines. Generally there is a mixture of the three functional groups in many resins: some, however, are predominantly of one species. This WBA resin classification is defined (similarly to the WAC resins) as not being capable of splitting neutral salts.

It should be mentioned that there are some resins which have mixed basicity, some weak and some strong, and these are called intermediate base anion exchange resins. The strongly basic groups are generally a minor part of the total capacity of these intermediate base anion exchange resins, with the major part being weakly basic groups. In the service cycle the WBA resins are mostly used following a SAC resin in the hydrogen form. This converts the cations in the water to hydrogen forms, which combine with the anions to form the corresponding acids. The acids would be carbonic acid, silicic acid, sulfuric acid, and hydrochloric acid. The WBA resins will only pick up the strong acids or mineral acids, with very little or none of the weak acids removed. If there is leakage of sodium from the SAC resin, this would be present as the neutral salt, sodium chloride, which would pass through the WBA resin unchanged.

The capacity of the WBA resins for removing mineral acidity is generally very high compared to the capacity of SBA resins, and the WBA resins are very efficiently regenerated. If the water quality required by the end use for the water does not require the removal of carbon dioxide or silica, the product water from the WBA resin will be adequatete. Some examples of where carbon dioxide and silica removal may not be required are process water, plating rinse water, bottled water, and low pressure boiler feed.

When treating waters high in mineral acidity and TDS, operating costs may be reduced by installing WBA resins following the primary SAC resin. The WBA resin removes the mineral acidity load and then an SBA resin removes the weak anions. This system can then use a portion of the waste caustic from the regeneration of the SBA resin to regenerate the WBA resin, thus lowering the chemical costs. (More of this in later discussions on various ion exchange resin systems in Chapter 7)

With WBA resin systems, the banding of sulfate on the top portion of the resin bed is followed by the somewhat more diffused layer of chloride. If there is any degree of strong basicity in the resin, there will initially be some removal of carbon dioxide and silica that will be rapidly displaced off the bottom of the resin bed as the service cycle progresses.

Chloride leakage will be the first sign of breakthrough, and this will usually follow or be accompanied by a drop in pH of the effluent and an increase in conductivity.

Another difference that should be noted in the way WBA resins perform the removal of acid from the water is that the entire acid molecule is removed on the weak base functional group of the WBA resin, and there is no true exchange of a hydroxyl (OH⁻) anion from the functional group on the resin, as is the case with SBA resins. The WBA resin acts more like an acid absorber, as shown by the following reaction equation:

$$RNH_2 \text{ (free base)} + HCl \rightarrow RNH_3Cl$$

As a result of this removal of FMA (free mineral acidity) from the water by the WBA resin, the pH will rise to 5 or 7, depending on the carbon dioxide content of the water. Because some WBA resins do have some strong base functionality, there are cases where the pH may rise to 8 or slightly higher in the first portion of the service cycle. Later, as carbon dioxide appears in the effluent of the WBA resin, the pH will drop to as low as 5.

There are a number of the WBA resins which swell considerably, by as much as 30% to 100%, in going from their fully regenerated free base form to their exhausted forms, while others have swelling of 8% to 20%. Most WBA resins swell in going from the regenerated to the exhausted form, but SBA resins swell in going from the exhausted form to the regenerated form. Provisions to handle these volume changes need be made in the consideration of equipment design, pressure drop problems, backwash bed expansion, and regeneration distributor placement.

Many of the WBA resins also have an ability to remove some organic contamination from the water, with the more porous resins showing the better removal capability. The use of WBA resins to remove organic contamination and protect the SBA resins that follow from fouling is important. Since the organics are not held as tightly on WBA resins as they are on the SBA resins, the WBA resins will release the organics more readily on regeneration.

Backwash Cycle

Following the service cycle there will normally be a backwash (BW) of the resin bed. This backwash cycle is of considerable importance to maintain the best operation of the resin bed for the following reasons:

- To expand the resin bed from its settled and sometimes packed condition that can result from swelling or shrinkage and particularly from a long service cycle at high flow rates.

- To clean the resin by flushing out the suspended dirt or particulate matter that may have been filtered from the feed water during the service cycle. Ion exchange beds are excellent filter media, because they are relatively small in particulate size and have ionic charges, similar to polyelectrolyte flocculants, to coagulate the extremely fine particles that may not have been removed by pretreatment filtration. The scrubbing action of the resin beads against each other during backwash helps to remove materials that may have coated the resin. Some of the dirt particles may be heavier than the low-density resins, but usually are much smaller in size, so they will be hydraulically washed off the resin. The faster the upward flow rate that can be maintained while still keeping the resin bed slightly below the top distributor takeoff, the more desirable it will be.

- To remove resin fines or broken resin particles. Fines are defined as those 50 mesh (0.297 mm or 0.0117 inches in diameter) or smaller. Resins do tend to break down physically over a period of time and use, and it is necessary to remove these resin fines to maintain good hydraulic conditions of the resin bed and to prevent channeling and pressure drop.

- To classify the resin bed with larger resin beads on the bottom and the finer resins on the top. This classification according to bead size provides the best conditions for uniform flow of regenerants, rinse flows, and service flow across the entire area of the resin bed.

Conditions for backwash vary with each type of resin because of their particle size distributions, differences in particle density, and possible difference in particle size. Charts with the recommended flow rates for various temperatures for each type of resin are available from resin manufacturers or equipment suppliers. Cation resins generally have higher densities than anion resins and therefore require higher superficial linear velocity to obtain the desirable resin bed expansion (40% to 50%) than do anion resins (70% to 100%). As has been pointed out earlier, careful attention to water temperature is necessary for the backwash cycle since the flow rate may need adjusting to maintain the resin bed expansion desired.

In the backwash of resin beds it is very desirable to maintain some back pressure by control of the flow on the outlet or discharge line. This prevents release of any air or gas from the backwash water in the resin bed, and thus prevents flotation of resin from the unit. Any gas released would attach in bubbles to the resin beads and cause clumps of resin beads to float. Even though they were good-sized beads, they would be lost out of the unit.

The time of the backwash will be dependent on the amount of debris present in the resin bed and the backwash should be continued until the discharge is clean and clear. Normally, 10 to 15 minutes is sufficient, but there might be extreme cases where 20 to 30 minutes may be required. If longer times than these appear necessary, special cleanup procedures might be considered.

The types of water used in the backwash cycle may be of importance. Such is the case with SBA resins, where use of hard water would cause precipitation of calcium and magnesium in the anion resin bed, due to the high pH of the incompletely exhausted resin bed. Soft or decationized (water from a SAC resin in the hydrogen form) water should be used.

Regeneration Cycle
Now we get into the real heart of the ion exchange operation, the chemical regeneration of the ion exchange resins, where variation in the resin, and control are of greatest importance to the quality and quantity of water that can be produced during the service cycle, as well as to the operation costs. Regeneration is the attempt to reverse the reaction that occurred in the service cycle. The main factors that must be considered are concentration, quantity of regenerant chemicals, contact time, and temperature. These factors will vary, depending on the type of resin, type of regenerant chemical, efficiency, and quality of the produced water desired. As in the other cycles thus far presented, each resin type will be covered.

We will discuss the co-current systems first and in later sections cover the contercurrent systems.

The first thing to remember is that the condition of the exhausted resin bed for all resin types is an ionically banded bed, with the most strongly held ions on the top portion of the bed, followed by bands of progressively less strongly held ions, and the most weakly held ions on the bottom of the bed. These bands may not be very distinct and are spread out relative to the quantities and types of various ions in the water being treated.

With co-current systems, the regenerant chemical first contacts the ion that is held most strongly on the top of the resin bed and this is usually with the regenerant chemical at its highest concentration and least contaminated form to most effectively remove the ion. (Later discussions will cover regenerant recycle where these conditions may be different.) Remember that the regenerant chemical is composed of a cation and an anion, and only one of the ions of the regenerant chemical exchanges for the ion that is removed from the resin's functional or exchange group. The displaced ion combines with the opposite ion in the regenerant solution.

An example would be in the regeneration of a SAC resin, used to soften water, and having been exhausted with calcium. The resin would be regenerated with sodium chloride brine. It would require two Na^+ monovalent ions to displace the Ca^{+2} divalent ion on the resin, with the Ca^{+2} combining with two Cl^- in the regenerant solution to form $Ca^{+2}Cl^-Cl^-$, or $CaCl_2$ as shown in the reaction:

$$RCa^{+2} + 2NaCl \rightleftharpoons R{\begin{matrix}Na^+\\|\\|\\Na^+\end{matrix}} + CaCl_2$$

The reaction does not go 100% to completion as the first volume of regenerant passes or contacts the resin beads on the top of the bed, and not all of the regenerant chemical in solution is used up on this first contact. Instead, this first quantity of regenerant, now at lower concentration and contaminated by the ions that have been removed or regenerated from the resin, goes on to the next layer of resin beads where it regenerates off some of the unwanted ions and becomes more contaminated in the process.

The first volume of regenerant solution finally approaches the point where the concentration of the regenerant chemical is too low and the contamination too high to do an effective job of pushing off the ions that need to be removed. In other words, the reaction is at the point where it is approaching equilibrium and will no longer move from left to right, as was shown in the reaction above. Now the arrow from left to right will be the same as the one from right to left.

If we look at the SAC resin that has been exhausted with both calcium and magnesium, the only differences that would be found in the regeneration would be that the magnesium band on the resin would be displaced by some of the calcium removed from the upper layer of the calcium, and the magnesium band displaced further down the column and eventually out to drain.

Looking at the succeeding volumes of regenerant solution coming down the resin bed, it can be visualized that the resin at the top of the bed would continue to progress toward complete regeneration. Depending on the amount or volumes of regenerant chemical that were used, the resin on the top will be the most completely regenerated, with progressively poorer regeneration occurring farther on down the column. The resin at the bottom would be the least regenerated of all.

From a practical and economic standpoint it is usually not possible to regenerate SAC or SBA resins to the degree where their total capacity is recovered. Generally the capacity being realized in large industrial systems will be in the range of 70%, but may range from 50% to 90%, with the top of the ranges costing considerably more for the added chemicals used.

It also should be realized that the degree of exhaustion or total capacity utilized in the service cycle will make a difference as to the amount of regenerant chemical that needs to be used in the next regeneration cycle to maintain capacity. For example, if the service cycle were cut short, or the ionic load in the water being treated were reduced with a fixed gallonage end point, then with the same amount of regenerant chemical being used, the next service cycle would be of better quality or greater capacity. On the other hand, if the run length were extended in the service cycle beyond the normal end point, or the ionic load were increased, the resin bed would be overexhausted and with the same amount of regenerant used in the next regeneration, the following service cycle would be shorter with poorer quality.

SAC resin (sodium form-water softening). With very few exceptions, sodium chloride is the regenerant chemical used for converting SAC resins from the exhausted form back to the sodium form. Other sodium salts that have been tried such as sodium sulfate (Na_2SO_4), sodium formate ($NaCHO_2$), and sodium tripolyphosphate ($Na_5P_3O_{10}$) have had very limited success. In the previous section we used the example of the sodium cycle regeneration of SAC resin, but there are other factors in the regeneration that need to be covered.

Salt quality. The quality of the salt used is important. Although it is not too generally recognized, there are numerous sources and qualities of salt, some of which have high amounts of contaminants that can cause problems in the regeneration of the resins. Where low hardness leakage (less than 1 ppm hardness) is desired, the use of high quality salt is required. Some rock salt and even some sea salt have calcium and magnesium contents, which when dissolved can produce saturated brine solutions with up to the solubility limits of 1,900 ppm of calcium (as ion) and 140 ppm of magnesium (as ion). Using this high hardness brine reduces the efficiency for removing the calcium and magnesium from the resin during the regeneration cycle, and results in higher leakage of hardness during the service cycle.

Another use for salt brine, where salt quality is important, is for the removal of organic foulants from SBA resins. If there is high hardness in the salt, the magnesium and calcium may precipitate when caustic is added to the salt brine to improve the organic removal, and the result may be further fouling of the SBA resin. If such fouling is encountered it would be necessary to use hydrochloric acid to remove these precipitates from the resin.

Seawater and brines from other sources can be used for regeneration of SAC resins for water softening applications, but the concentration and purity of these brines must be considered with regard to their utility and efficiency.

Brine concentration. It is reasonable to expect that the higher the brine concentration the greater will be the driving force to push the hardness off the ion exchange groups on the resin and replace or exchange them with sodium. This is a reasonable expectation, but there are some potential problems that can be encountered. The highest concentration of sodium chloride that can be obtained at normal temperatures is approximately 26% by weight. (Most often, if there is a rapid use of brine and therefore a short time for the salt to dissolve in the water, the brine concentration that can be reached may be no more than 23% to 24%.)

A high-concentration salt solution tends to dehydrate (remove water from) any substance that has water in it with a lower salt content. The substance with the higher water concentration (or lower salt concentration) will try to pass water to the solution with the lower water content (and higher salt content) so as to equalize the salt concentration in each. This process is called osmosis. The greater the difference in salt content, the greater the pressure against any permeable membrane (such as the solid surface of the ion exchange resin bead and the liquid water solution of salt) by the water attempting to pass to the other side and dilute the high salt content solution.

As the dehydration of the resin occurs at the outer skin of the resin bead first, the surface of the bead will shrink (or try to shrink). Like the drying of a mud flat after a long drought, the surface will crack as the water is removed, and the resin beads can split or crack. This phenomenon is called osmotic shock. Some of the resins are more resistant to osmotic shock than others, with the macroporous resins being generally more resistant than the gellular resins (both anion and cation). The breakdown of the gellular resins is generally not great, however, and some grades hold up quite well, with only slight loss occurring over extended exposure to cycling with saturated brine.

There is also some shock experienced with the rehydration of the resin beads after brining, when the resin is rinsed to remove the brine. In this case the resin beads swell as the salt is rinsed from the beads, causing another physical strain on the beads. In spite of the potential problem of osmotic shock with concentrated brine regeneration, it is used when the maximum capacity is required and resin replacement costs can be justified.

Reducing the concentration of the brine does reduce the osmotic shock problem and in the normally used concentration range of 8% to 12%, there is very little problem with reduced resin life. Most industrial and domestic water softeners are designed to operate in these dilute brine concentration ranges.

Another reason for dilution of brine is to have a sufficient volume of regenerant solution so as to uniformly contact all of the resin with the distribution system used, particularly at the lower, more efficient regenerant levels. Contact time of the regenerant with the resin of at least 30 minutes is desirable. The contact time is calculated from the time the regenerant is introduced into the resin bed until it is displaced from the resin bed by the displacement rinse.

At an efficient level of 6 pounds of salt per CF of SAC resin for industrial softening, the gallonage of saturated brine used would be about 2.4 gallons (since there are approximately 2.5 pounds of salt per gallon of saturated brine). The minimum displacement rinse would be one void volume, or approximately 40% of one CF (7.5 gallons per CF) or $0.4 \times 7.5 = 3$ gallons. The 2.4 gallons of saturated brine plus 3 gallons of displacement rinse would be a total of 5.4 gallons. With a contact time of 30 minutes the flow rate per CF would be 5.4 gal/30 min = 0.18 gpm/CF. Flow rates much below 0.25

gpm/CF are not normally recommended. At this low flow rate it is difficult to obtain good distribution in large-diameter vessels, and since the concentrated brine is of higher density than water, it may more rapidly take the course of least resistance and channel through the resin bed, not making good contact with all the resin.

Looking at the low end of the brine concentration curve, it would be expected that there would be reduced capacity and regeneration efficiency when the brine concentration gets too low. This would occur because the driving force to displace the hardness with sodium would be low. The drop-off in capacity with brine concentrations below 10% is more rapid as the concentration decreases. At 2% brine concentration, the capacity would be approximately 90% of what it would be at 10% brine concentration. One might also expect higher hardness leakage during the service cycle with the use of very dilute brine.

Another example of the effect of low concentration brines on the capacity obtained is in the use of seawater or low concentration salt-well brines in the regeneration cycle. With seawater of 2.5% to 2.75% salt concentration, the capacity obtained would be only 60% to 70% of the capacity that could be obtained by use of 10% low-hardness brine. Part of the sacrifice in capacity would be due to the high hardness (approximately 5,500 ppm as $CaCO_3$) of seawater, which would reduce the regeneration efficiency and increase the leakage in the service cycle. However, it is possible to obtain some capacity by using brines with salt concentrations as low as 0.75%, but capacity of only about 30% of that obtained at 10% concentration would be expected, again with higher hardness leakage.

Regeneration Level. The amount of regenerant used will of course have an effect on the amount of capacity that can be obtained, up to the total capacity available for the resin being used. Each resin has a given amount of capacity built into it, based on the amount of ion exchange sites that can be put on or into the resin structure and how available those exchange sites are to the ions in solution. This total capacity can be determined by standard testing procedures, as described in most resin manufacturers' literature.

In the use of SAC resins for water softening, it is not practical or economical to try to totally regenerate the resin to the sodium form. New resin, as received from the resin manufacturer, is normally supplied completely in the sodium form, and the first service cycle will produce maximum capacity. Subsequent cycling will never recover that capacity without heroic amounts of regenerant. A compromise must therefore be reached on the amount of salt (or regenerant level) that is to be used to obtain the best capacity per pound of regenerant used and to maintain an acceptable quality of produced water in terms of leakage.

In water softening with the standard high-capacity types of SAC resins, for industrial applications and for most waters, this level is usually in the range of 6 pounds of salt per cubic foot of resin (or 6 lbs/CF) to obtain 22 Kgr/CF capacity. In some home water softening applications the range may be 8 to 15 lbs/CF to obtain 24 to 32 Kgr/CF capacity. In some municipal softening applications where they can (and usually do) tolerate some hardness leakage (in fact where it is desirable to maintain some hardness in the water to prevent corrosion of the distribution piping system), a salt level of 5 lbs/CF has been found satisfactory to obtain capacities in the range of 23 Kgr/CF.

Another way to show efficiency is in terms of pounds of salt per kilograin of resin capacity obtained, expressed as $CaCO_3$. For industrial softening this would be approximately 0.27 lbs/Kgr; for domestic it would be 0.33 lbs/Kgr to 0.47 lbs/Kgr; and for municipal softening, where high hardness is acceptable, 0.22 lbs/Kgr.

It must be remembered that the hardness leakage that can be tolerated, the quality of the water to be treated, the concentration of the regenerant, and the temperature all can influence the efficiency that can be realized from a given amount of regenerant or regeneration level.

SAC Resin—Hydrogen Form. As in the case of the sodium cycle regeneration example, we must look at the condition of the exhausted resin bed with bands of calcium, magnesium, sodium, and of course the smaller remaining band of some resin left in the hydrogen form.

In the regeneration of the resin to the hydrogen form compared to regeneration of the resin to the sodium form—even though a hydrogen ion (H^+) and sodium ion (Na^+) are each monovalent—sodium is held more tightly by the resin than is hydrogen. This means that to shift the reaction for replacing the unwanted ions from the exhausted resin with hydrogen will require greater effort than for sodium. This becomes more apparent when looking at the efficiencies that are possible in hydrogen cycle regeneration as compared to those for sodium cycle regeneration, and considering the equivalent weight differences between the chemicals used.

The two chemical regenerants used in the hydrogen cycle operation are hydrochloric acid (HCl) and sulfuric acid (H_2SO_4), with the latter being more commonly used because of cost considerations. Hydrochloric acid will be discussed first as it is less complicated in its use than sulfuric acid. Unfortunately, however, hydrochloric acid is more expensive, and it causes some handling problems in storage and corrosion.

Hydrochloric acid is available normally as a 20° Baumé (32% concentration) solution. Most commercial grades of hydrochloric acid are acceptable for cation resin regeneration.

Hydrochloric Acid Concentration. There are no cations that would be expected in water supplies that would form insoluble precipitates when regenerated from the resin with hydrochloric acid (as there are when sulfuric acid is used). Thus concentrations as high as 10% can be used. However, the best concentration to be used will depend upon the feedwater analysis as to the percentage of divalent cations (calcium and magnesium) as compared to the percent sodium (or other monovalent cations). Since the divalent cations are held more strongly than the monovalent cations, the higher the concentration of hydrochloric acid, the better and more completely they will be removed. Using the same reasons for not using very high concentrations of regenerant as was discussed for brine regeneration, 8% to 10% hydrochloric acid would be best used when divalent cation concentrations are high, while better capacities will be obtained with 2% to 5% acid when sodium concentrations are high. A compromise in the acid concentration can be made for waters of intermediate divalent to monovalent ratios. One also should consider the time factor, as with lower concentrations larger volumes of regenerant will be used and there will be larger volumes of waste.

With hydrochloric acid, the contact time should be at least 30 minutes, but to improve the removal of calcium and magnesium, longer contact time is beneficial and will show up as improvement in capacity. Divalent cations, particularly calcium, are strongly held by the resin, and what is not removed during the regeneration cycle will remain on the exchange groups during the subsequent service cycle, with very little or no leakage evidenced (depending on the feedwater); but the amount on the resin will reduce the amount of exchange groups available and therefore reduce the usable capacity of the resin.

Sulfuric Acid Concentration. Sulfuric acid is generally available at lower cost and higher concentration (96% to 98%) than hydrochloric acid. It requires less storage volume and less freight costs, and it can be stored in steel tanks. For these reasons it is the acid of choice for most industrial plants. Observe this caution in the dilution of concentrated sulfuric acid: Never add water to the acid in an open tank! Always add the concentrated sulfuric acid to the water with good mixing to prevent spattering and dangerous safety conditions. The heat of solution (heat generated when concentrated acid is added to water) is considerable, and can be high enough to cause deformation of thermoplastic pipes or parts.

All safety precautions and procedures should be observed when handling any of the corrosive acids and caustic chemicals used in the regeneration of ion exchange resins. Nitric acid is seldom used as a regenerant in water treatment applications, but nitric acid and other strong oxidizing agents can cause explosive reactions when mixed with organic ion exchange resins, and knowledgeable persons should be consulted before handling such materials.

Chapter 6

The concentration of sulfuric acid used as a regenerant for SAC resins is of greatest concern when dealing with waters containing calcium, in view of the potential for precipitation of the relatively insoluble calcium sulfate ($CaSO_4$) or gypsum ($CaSO_4 \cdot 2\ H_2O$) on or in the resin or the waste system. The solubility of calcium sulfate is approximately 2,500 ppm under normal conditions, but supersaturated (concentrations higher than the solubility limit) solutions can exist for some period of time (before precipitation occurs), depending on temperature and sulfuric acid concentration.

Keeping temperature below 85°F, the concentration of sulfuric acid no higher than 2%, and the flowrate as high as 2 gpm/CF will usually allow the regenerant solution to get through the entire resin bed and out to waste before precipitation can take place.

The calcium percentage of total cations in the treated water supply will determine the depth of the calcium band formed in the resin bed; this percentage therefore determines how much calcium will have to be swept out of the bed (with the acid at low concentration) to keep below the level that would precipitate. It also determines how soon the acid concentration can be increased to get more of the calcium out of the resin. In order to get good removal of most of the calcium and obtain high capacity, the acid concentration must be increased in the later stages of regeneration. The higher the ratio of calcium to total cations, the more steps (of increasing the concentration of the sulfuric acid and reducing the flowrate) may be required to finally arrive at a high enough concentration to remove most of the calcium without the dangers of forming calcium sulfate in the resin bed. From experience, the following "rules of thumb" charts may be helpful in reducing the probability of calcium sulfate precipitation. (See Table 6-A).

For practical reasons, many equipment engineering firms try to reduce the complexity and cost of acid-feed systems to two stages of concentration by extending the low concentration acid to approximately one half of the total pounds of acid, and using the second half at the higher concentration. This practice has worked well with little or no problems of $CaSO_4$ precipitation or capacity sacrifice, particularly in the normal ranges of percent calcium of total cations and regeneration levels.

It is well to reemphasize the point that calcium sulfate precipitation is a time-, concentration-, and temperature-influenced problem, and that it is necessary to get the supersaturated calcium sulfate solution out of the resin bed and distributor systems before precipitation can occur. Calcium sulfate crystals will form on any surface, such as tank walls, resin beads, distributor screens, slots, or pipes; will result in calcium leakage in the service cycle; and will require an involved cleanup procedure (usually with hydrochloric acid) to get the system back in proper operation.

One way of checking to see if the stepwise or progressive regeneration

TABLE 6-A
Conditions to Reduce Probability of CaSO₄ Precipitation with Sulfuric Acid Regeneration

% Ca of total cations	Flowrates gpm/CF
0	0.5
25	0.5
50	1.0
75	1.5
100	2.0

Regeneration level total pounds per CF	Pounds used at each concentration of H_2SO_4			
	2%	4%	6%	8%
3	3			
4	2	2		
5	2	3		
6	2	2	2	
7	2	2	3	
8	2	3	3	
9	2	2	2	3
10	2	2	3	3

programs are properly set is to take samples, in clear glass bottles, of the waste regenerant immediately as it comes from the resin tank, throughout the regenerant introduction step, and into the displacement rinse. Immediately, and after 2, 5, and 10 minutes, observe the clarity of the samples for tendency to form precipitates. The samples, of course, should be clear as they are taken and should remain clear for sufficient time for the waste to be disposed into the waste handling system.

It should also be recognized that apparently only one hydrogen of the sulfuric acid (H_2SO_4 or $H^+ HSO_4^-$) is utilized when divalent cations are being removed from the cation resin in the regeneration process, making the efficiency of sulfuric acid less than that of hydrochloric acid, with resulting lower capacity where high hardness waters are being treated.

Regeneration Level. In most of the discussions of regeneration level, the quantity of regenerant is given in pounds of the chemical at 100% concentration and not as supplied (HCl is usually supplied at 32%, H_2SO_4 at 96% to 98%, and NaOH at 50%), unless otherwise stated. In most of the applications for hydrogen-form SAC resin, the requirements for performance are based on leakage of sodium allowable (in the uses for dealkalization, sodium leakage is not of main concern, so the most efficient regenerant levels are used).

Before going further, it would be well to review some of the reasons for sodium leakage during the service cycle as being affected by the regeneration cycle. Since it is not economical to use enough regenerant in the regeneration cycle to completely remove all of the contaminating ions from the exhausted resin, there will be a band of contaminating ions (Ca, Mg, and Na) left on the bottom of the resin bed (in co-current operation). The calcium and magnesium that are not removed during the regeneration cycle are more strongly held on the resin exchange sites than is sodium, and as a result will not tend to leak off the resin during the following service cycle in the same manner as sodium. (They will, however, reduce the capacity by the amount remaining on the resin exchange sites.) The sodium leakage at the beginning of the service cycle is a result of sodium being left on the bottom of the resin bed after an incomplete removal of sodium during the regeneration, and then being eluted off the resin by the FMA formed in the well-regenerated upper portion of the SAC resin bed. The sodium leakage during the service cycle could therefore be illustrated as follows:

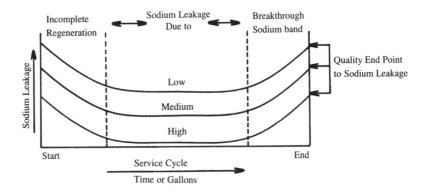

Figure 6-4. Sodium leakage profile.

The sodium leakage is not only dependent on the regeneration level, but also on the quality of the water to be treated, the regeneration flowrates, and the concentration of the regenerant, all of which have an influence on the efficiency of the amount of regenerant used.

In hydrogen-form SAC resin service, the quality of the water (or water analysis) has to be considered more carefully than in the sodium-form softening application because of the effect that the percentage of alkalinity, calcium, magnesium, and sodium have on the capacity, as well as the sodium leakage at a given regenerant level.

Percent sodium. The higher the percentage of sodium of the total cations in the water analysis, the greater will be the depth of the sodium band formed during the service cycle. Although sodium will be rapidly removed during the early portion of the regeneration cycle, the more sodium there is initially on the resin bed, the more will be left on the bottom of the bed during the regeneration cycle. The result will be higher sodium leakage during the service cycle. Higher regeneration levels will have to be used on high-percentage-sodium waters to meet sodium leakage levels than would be needed on low-percentage-sodium waters with lower regeneration levels. Typical curves of sodium leakage related to percent sodium of total cations at a set of regeneration levels and fixed alkalinity would be illustrated thus:

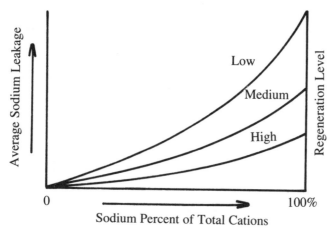

Figure 6-5. Sodium leakage as a function of sodium ratio.

Alkalinity. The alkalinity of the water to be treated as percent of total anions will also have an influence on the sodium leakage during the service cycle and therefore on the regeneration level that must be used to meet a desired leakage limit. The reason for this is that in waters with high mineral acidity (Cl and SO_4) and low alkalinity (HCO_3), the result of replacing the cations Ca^{+2}, Mg^{+2}, and Na^+ with hydrogen (H^+) is the formation of the corresponding acids of the anions (HCl and H_2SO_4). These are strong free mineral acids (FMA) and can elute off (or regenerate off) the sodium left on the bottom of the resin bed to give high sodium leakage.

With high alkalinity in the water to be treated, the alkalinity combining with the hydrogen ion will form carbonic acid (H_2CO_3), which is too weak an acid to cause sodium to be removed from the resin bed and have it show up as sodium leakage.

The shape of the curves at different acid regeneration levels showing the effect of average sodium leakage with varying alkalinity percentage of total anions (with sodium level constant) would be roughly as illustrated below:

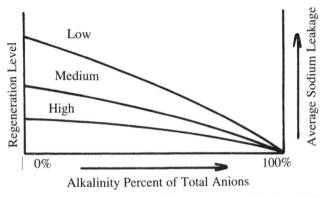

Figure 6-6. Sodium leakage dependency on regeneration level and alkalinity.

From the above it can be seen that if sodium leakage is the criteria or factor of quality required in the water produced from the SAC resin in the service cycle, the sodium and alkalinity levels in the water to be treated will determine the acid level that must be used in the regeneration cycle to achieve this goal.

With the detailed data available from the resin manufacturers on the particular SAC resin to be used, the acid regenerant level can be determined, as well as what capacity can be expected for that particular water analysis.

Capacity. The capacity that can be obtained is, of course, dependent on the acid that is used. With hydrochloric acid, higher capacities are possible than with sulfuric acid. This is primarily due to the fact that even though sulfuric acid has two hydrogens compared to one for hydrochloric acid, it appears that only one hydrogen ion is utilized effectively with sulfuric acid. The problems involved with the calcium sulfate precipitation require use of lower concentrations of sulfuric acid, which results in lower capacities achieved per pound of acid used.

Magnesium was mentioned as another variable which can influence the capacity that can be obtained. This is primarily due to the fact that as the magnesium-to-calcium ratio increases, there are less problems with calcium sulfate precipitation (as magnesium sulfate is quite soluble). Higher acid concentration can be used to improve removal of the divalent ions and as a result higher capacities can be obtained.

Capacity curves for varying regeneration levels could be illustrated by a family of curves. These would also depend on the sodium leakage allowable as determined by the composition of the water being treated for both percent sodium and percent alkalinity as shown below:

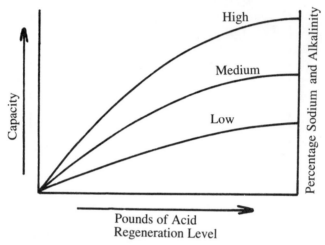

Figure 6-7. Capacity dependency on regeneration level and water analysis.

Regeneration Efficiency. Regeneration efficiency of hydrogen-form SAC resins is dependent on the regeneration level to meet the sodium leakage requirements, particularly in co-current regeneration, and therefore cannot be a matter of choice. For sulfuric acid, the regeneration efficiency in pounds per kilograin (lbs/Kgr) of cations removed can range from an approximate low of 0.25 lbs/Kgr to a high of 0.50 lbs/Kgr. With hydrochloric acid, regeneration efficiencies would be in the approximate range of 0.15 lbs/Kgr to 0.30 lbs/Kgr.

Where sodium leakage is of no great concern, then the lower regeneration levels would be the obvious choice for low cost of operation; however, with the lower capacities there is an increase in capital costs, since more resin and larger tank size will be needed to give the desired service cycle.

WAC Resins. As indicated in the discussions of WAC resins in the service cycle, calcium and magnesium cations are more tightly held than is the case with the SAC resins, as long as the pH is not too low or acidic (this would be the case where the alkalinity of the water was significant as compared to the chlorides and sulfates). The exhausted resin operating in the hydrogen form would therefore have the calcium in the top band followed by magnesium and finally sodium (if the resin were not exhausted to the hardness end point). Because of the much greater affinity of the WAC resins for divalent cations over that of the monovalent sodium ion, it can be recognized that the WAC resins cannot be regenerated by sodium chloride when the WAC resins have been used in the softening process, as are the SAC resins. So, for all uses of WAC resins for either sodium-form softening or hydrogen-form decationizing, the first step is to convert the resin to the hydrogen form, which is the form most favored by the carboxylic (COOH) exchange site on the resin.

The general choices of acids are those used for the SAC resins, hydrochloric acid and sulfuric acid (and sometimes the waste acid regenerant from the SAC resins is used to regenerate the WAC resins in the same system). The choice of acid favors hydrochloric acid in order to get around the calcium sulfate precipitation problem that could occur when sulfuric acid is used. This choice becomes the odds-on favorite with WAC resins because of the higher calcium concentration on the resin (higher total capacity); the slower diffusion of calcium and magnesium ions out of the resin bead (mostly due to the shrinkage of the WAC resin beads as they go from the exhausted to the regenerated form); and the volumes of waste when using the low concentration of sulfuric acid required to get around the calcium sulfate precipitation problem.

The difference in regenerating WAC resins as compared to the SAC resins is that with SAC resins the amount and concentration of acids to convert the resin from the exhausted form to the hydrogen form is greater because of the greater attraction for other cations than for hydrogen, while the reverse is true for the WAC resin. Because of this attraction for hydrogen over other cations by the WAC resins, the regeneration can be accomplished with almost the equivalent amount of acid to the amount of other cations on the resin. This is further shown by the fact that both hydrogens of sulfuric acid are utilized in the regeneration of WAC resins, and even a partial regeneration can be accomplished with a very weak acid such as carbonic acid (CO_2 in water under pressure).

Hydrochloric Acid Concentration. As discussed above, the regeneration of the WAC resins to the hydrogen form does not require high concentration of acid to remove even the divalent cations from the resin, so concentrations of acid as low as 0.5% could be used, but of course the waste volume would increase.

Because of the excellent efficiency of regeneration, low regeneration levels are used. If the acid concentration were too high, the total regeneration volume would be very low and the flowrates used to obtain reasonable contact time would be low, creating possible distribution problems. With hydrochloric acid, concentrations of 4% to 5% are normally used with good results.

Sulfuric Acid Concentration. As discussed under the SAC resin regeneration section, the calcium sulfate precipitation problem is of considerable concern, and with WAC resins, there is reason for even greater concern because of the more rapid exchange of calcium with hydrogen (therefore increasing the concentration of calcium in solution) and the higher amount of calcium (due to the higher capacity of WAC resin as compared to SAC resin for calcium) per unit volume of resin.

A further problem with WAC resins and calcium sulfate precipitation is the shrinkage of the resin in going from the exhausted form to the regenerated or hydrogen form. Since the shrinkage first occurs on the outer shell of the resin bead where the resin is first contacted with the acid, it makes it more difficult for the calcium ions to diffuse out, and results in a higher concentration of calcium in the presence of sulfate ion and a potential for calcium sulfate to precipitate within the resin bead.

It is for these reasons that it is necessary to go to concentrations as low as 0.3% to 0.8%, depending on the flowrates used. The higher the flowrate, the faster the calcium sulfate is flushed from the resin bed before precipitation can occur. Flowrates of 0.5 gpm/CF at the lower concentrations, to 2 gpm/CF at the higher concentrations, are used to obtain satisfactory regeneration and to overcome the calcium sulfate precipitation problems.

Converting to Sodium Form for Softening. As indicated previously, the WAC resins in the sodium form have application in softening of high TDS water, and even brines. The exhausted resin in the calcium and magnesium form can not be directly converted to the sodium form with sodium chloride brine, but has to be first regenerated with acid to the hydrogen form and then converted to the sodium form by sodium hydroxide (NaOH) or a salt of sodium with a weak acid such as sodium bicarbonate ($NaHCO_3$), or sodium carbonate (Na_2CO_3).

A direct conversion of the resin from the exhausted form to the sodium form can be made by a sodium citrate solution in combination with sodium chloride and sodium hydroxide. This would be more expensive than the use of acid/caustic regeneration and would result in slightly lower capacities, but would eliminate the use of corrosive and difficult-to-handle acids.

When converting the WAC resins to the hydrogen form, the same procedures would apply as indicated for the other applications. The additional step of converting the resin from the hydrogen form is accomplished by using a 4% solution of sodium hydroxide (after rinsing out the excess acid). Where the capacity is fully utilized, the acid level should be in the range of 7 to 10 lbs/CF of 100% HCl, and the caustic level should be at approximately 7 lbs/CF, depending on the specific resin used.

Regeneration Level and Efficiency. With WAC resins, the regeneration level is primarily determined by the capacity utilized in the service cycle since the regeneration is very efficient, with the resin preferring the hydrogen form over most any other form. Capacities as high as 60 Kg/CF are possible to achieve and at this level at 100% efficiency would require 6.3 lbs/CF (as 100% HCl) of hydrochloric acid or 8.7 lbs/CF (as 100% H_2SO_4) of sulfuric acid. The regeneration efficiency at this level would be 0.105 lbs/KG for HCl and 0.143 lbs/Kg for sulfuric acid.

With some expected losses in efficiencies in commercial and industrial operations, and to insure complete regeneration, 110% to 120% of equivalent acid would be required. This would put the regeneration levels normally used in the range of 6.9 to 7.6 lbs/CF for HCl and 9.6 to 10.4 lbs/CF for H_2SO_4. When the lower levels of capacity are used, then it is only necessary to use a regeneration level that would correspond to 120% of the capacity used. An example would be if only 25% of the total capacity of 60 Kg/CF were used (or 15 Kg/CF), then 25% of the 7.6 lbs/CF HCl (used for the 60 Kg/CF), or 1.9 lbs/CF could be used to obtain the lower capacity of 15 Kg/CF.

The applications for WAC resin where the ultimate high capacity can be achieved are few, and most of the applications of these resins are where the lower capacities are applicable and the major advantages to be achieved are regeneration efficiency and selectivity not possible with SAC resins.

SBA Resins. In the discussion of the service cycle of SBA resins it was pointed out that the anions are loaded onto the resin bed from top to bottom in the order corresponding to how strongly the particular anions (SO_4, Cl, HCO_3, and SiO_2) were held, and that there are a number of differences in regeneration of SBA resins (as compared to SAC resins) because of silica and its special characteristics.

The regeneration chemical used for SBA resins is sodium hydroxide (caustic), which is usually obtained in the 50% solution and should be maintained at temperatures above 52°F or it will solidify. Where SBA resins are used for DI systems, there are no other readily available strongly basic regenerants used, although potassium hydroxide (KOH) and calcium hydroxide ($Ca(OH)_2$) have been used in a very few cases.

Quality of Caustic. The quality of sodium hydroxide or caustic used is of greater concern than for most other regenerant chemicals for ion exchange resins because of the impurities present in some grades that can readily affect performance of the SBA resins. The impurities that should be monitored are iron, sodium chloride, sodium carbonate, and sodium chlorates.

Iron can be a problem, as the iron hydroxide present can physically precipitate on the resin as ferric hydroxide, $Fe(OH)_3$; and since the SBA resin is not exposed to acids or low pH in normal operation, the iron will continue to accumulate and will eventually reduce the available capacity by fouling.

Sodium chloride present in the caustic reduces the effective sodium hydroxide available for regeneration and possibly would leave a small amount of residual chloride on the resin, resulting in very low chloride leakage. Sodium carbonate would be another factor in reducing the effective sodium hydroxide available.

Sodium chlorate is objectionable since the chlorate is bound by the exchange groups on the SBA resins by a factor of approximately ten times that of chloride and therefore will reduce the available capacity of the resin accordingly. Although chlorate can be a strong oxidant at pH below 7, above a pH of 7 there is no great hazard for oxidation of the resin or its exchange sites. With SBA resins the pH would not get below 7, but with WBA resins the pH of the exhausted resin can readily be on the acidic side and oxidation of the resin or exchange sites can occur.

Silica. Removal of sulfate, chloride, and bicarbonate anions is fairly straightforward, except in a few specific high quality applications. For most applications of SBA resins, silica is the quality control factor around which the regeneration cycle must be designed.

Silica has unusual properties as compared to the other anions in both the service cycle and the regeneration cycle, due to its tendency to polymerize on or in the resin structure, particularly with time. The longer the time between regenerations (long service cycles), the greater is the tendency for polymerization to take place.

In order to get all of the silica off the SBA resin and to solubilize the polymeric silica, which can be done with the caustic regenerant, the two most important factors are time (to solubilize the polymeric silica) and temperature (the higher the temperature of the caustic solution, the faster the polymeric silica will go into solution).

If all the silica is not removed during the regeneration cycle, there will be higher silica leakage in the following service cycle. The silica not removed will be located near the bottom of the resin bed and will be leached off the resin by the high pH created by whatever small amount of sodium leakage there is from the cation resin bed preceding the SBA resin.

Caustic (NaOH) Concentration. In regeneration of the SBA resin with caustic, there are no great problems in regenerating most of the other anions from the resin under regeneration conditions satisfactory for removing silica. The concentration of caustic at 4% has been found to be a good compromise by having sufficient driving power to remove most anions from the resin, by providing sufficient volume of regenerant to contact all the resin at a flowrate acceptable to give a minimum contact time of 60 minutes, and by not causing overly high osmotic shock to resins which have about 20% expansion as they go from the exhausted form to the regenerated form.

Lower concentrations, down to 2%, can advantageously be used with low regeneration levels to provide longer contact time and to increase efficiencies and obtain good silica removal. Higher concentrations do not greatly improve regeneration efficiencies and do increase probabilities of osmotic shock. It might be difficult with high concentrations to control contact time and at flowrates that would allow good uniform distribution with a high density regenerant (close to the density of the SBA resin to possibly cause resin flotation).

Regeneration Level. For most applications of SBA resins, it is desired that the weak anions of bicarbonate and silica in the water are removed as well as the FMA (or one could use WBA resins for the FMA removal with higher capacities and better regeneration efficiencies). With the leakage of silica occurring before bicarbonate leakage, silica leakage is the controlling factor in determining the regeneration level and other operating procedures, such as contact time and temperature of regenerant, that influence the silica leakage.

Silica Leakage. The other factors influencing the silica leakage, which must be considered when choosing the regenerant level, are the sodium leakage from the preceding SAC resin, the percentage of silica to total anions, temperature of the water to be treated, and flowrates used in the service cycle. To illustrate these silica leakage factors, the following graphs (Figure 6-8a-8f) roughly show these influences.

It can be seen from these graphs that besides the regeneration level only Figure 6-8c (Regenerant Temperature) and Figure 6-8e (Regenerant Contact Time) show influences in silica leakage that can be controlled in the regeneration cycle.

The other factors (water temperature, silica to total anions, sodium leakage from the SAC resin treatment of the water, and service cycle flowrate) that influence the silica leakage are all based on the water to be treated, and to a great degree are not controllable when the regeneration procedures are being calculated. Of these factors, the percent silica of total anions and sodium leakage from the SAC resin are most important, since service water temperature normally is not outside acceptable limits, and flowrates can be controlled by volume of SBA resin that can be used.

It is therefore necessary to utilize the data available from the ion exchange resin manufacturers to come up with the silica leakage values that are possible at various regeneration levels and to give acceptable silica leakage limits. High regeneration temperature and long contact time are most important in achieving complete removal of the silica and thereby reducing the silica leakage that is evident in these graphical representations.

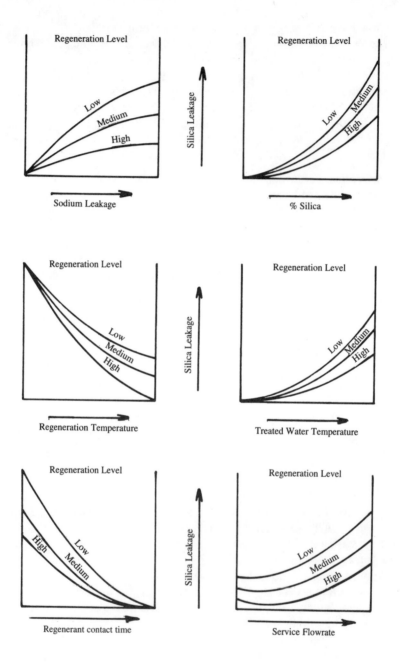

Figure 6-8. Effect of operation conditions and regeneration on silica leakage.

There are, of course, limitations on how high a temperature can be used on SBA resins, since the exchange group used to give the strongly basic characteristics of these resins is a quaternary ammonium group which has limited resistance to high temperatures. The regenerated hydroxide (OH) form of the quaternary ammonium group (for short called "quat") is the least resistant to high temperatures for any extended period of time. Some breakdown of the exchange group will occur to reduce its strongly basic properties in the regeneration process if the temperature is too high. For the type I SBA resins the limit is 120°F, while with the type II SBA resins this limiting temperature is 105°F. The acrylic SBA resins are usually limited to 95° to 100°F.

It must be understood that these temperature limits are not a precise limit below which there is no effect and above which there is immediate or complete breakdown, but that the time span for which the resins are exposed to the temperature is also a factor. Short-term exposure to even 130°F for type I or 120°F for type II SBA resins will not cause dramatic loss of capacity, but should be limited. In the exhausted form, the SBA resins have better stability to endure high temperatures and can stand exposure of up to 170°F.

It would therefore be desirable to use the highest temperature permissible in the regeneration of the SBA resin to achieve the lowest silica leakage, and the longer the temperature can be maintained at the maximum, the lower the silica leakage will be. With this goal in mind and since the exhausted resin is at ambient temperature, or temperature of the water used in the service cycle and backwash step, it would improve the silica regeneration if the temperature of the resin could be raised prior to the introduction of the dilute caustic, which is preheated to the maximum temperature allowable. In fact, just by the preheating of the resin bed with hot softened or DI water, there can be observed in the effluent a measurable amount of silica coming from the resin bed even before the hot caustic regenerant contacts the resin. Continuing the hot water into the displacement rinse will continue to improve the silica removal.

Contact time of the dilute caustic regenerating solution with the resin should be as long as possible in order to give time for dissolving the silica that has polymerized in the resin bed. A contact time of 60 minutes (usually obtained with regeneration flowrate of about 0.25 gpm) has been found to be a practical amount of time to achieve good regeneration under most situations. If silica leakage cannot be achieved because of the lack of hot dilution water for the caustic, reducing the flowrate to give longer contact time or even reducing the concentration of caustic while maintaining the flowrate will result in the longer contact time desired and will improve the regeneration.

Capacity. Regeneration levels used can generally be in the range of from 4 to 10 pounds per CF (for co-current regeneration systems). The capacities that can be obtained at these levels vary widely with the silica percentage of total anions, as well as the alkalinity or bicarbonate (HCO_3) percent of total anions (being the other weak anion), and chlorides percentage of FMA,

$$\frac{\text{chlorides}}{\text{chlorides} + \text{sulfates}} \times 100.$$

The reason why chlorides influence capacity is that this monovalent anion is not held as tightly by the SBA resin as is the sulfate, and whatever chloride remains on the resin bed after regeneration would tend to occupy exchange sites that would not be available to the weak anions of bicarbonate and silica and would cause early leakage of the silica. The higher the chloride percentage of FMA, the lower the capacity. Generally chloride levels below 10% to 20% of the FMA have little effect on the rated capacities (as obtained from resin manufacturers' published data).

Regeneration Efficiency. As is the case for most ion exchange resins, and probably more so for the SBA resins, it is difficult to express a capacity or efficiency figure except for an "average case" water analysis. Just to give a rough idea of regeneration efficiencies for SBA resins, they would be in the neighborhood of 0.35 lbs/Kg for type I and 0.23 lbs/Kg for type II, with capacities ranging from 12 to 17 Kg/CF for type I and 17 to 22 Kg/CF for type II.

WBA Resins. As is the case with WAC resins, the regeneration of WBA resins is accomplished with a minor excess of regenerant, and utilization of near-total capacity can be achieved.

Regenerant chemicals that can be used for WBA resin include weakly basic ammonia (NH_3) or ammonium hydroxide (NH_4OH), as well as sodium carbonate (soda ash Na_2CO_3), and sodium hydroxide (NaOH). Cost, availablity, effects on rinse characteristics, and type of systems in which the WBA resins are used will mainly determine which of the regenerant chemicals is chosen.

Regenerant Concentration. There are no significant problems in regeneration of WBA resins as far as precipitation or insoluble salts being formed with any of the regenerants normally used, and choice of regenerant concentrations are generally chosen to obtain the best compromise of regenerant volume and contact time. Low concentrations can be used, but this will increase the waste volumes accordingly. Concentrations from 2% to 8% have been utilized, but the most common concentration used and recommended is 4% for any of the regeneration chemicals used.

There are a few of the WBA resins which do have some temperature limitations, both on regeneration and service cycle, of 104°F, but most of the others have capabilities of operating up to 140° to 212°F. The temperature factor will have some influence on resin choice, depending on the requirements of the service cycle and the regeneration cycle, where the waste heated caustic from the regeneration of a SBA resin may be used for regenerating the WBA resin in the system.

Contact time of the regenerant on the WBA resin is not of too great a concern, because of the favorable conditions to change the resin from the acid form to the free base form, as shown in the following reaction using the HCl-exhausted form of the resin as an example:

$$RNH_3^+Cl^- + NaOH \rightarrow RNH_2 + NaCl + H_2O$$

Because the shrinkage of some WBA resins in going from the exhausted form to the free base form, there is some need to allow for time of migration of the regenerant to the exchange sites in the resin bead and for the waste products to migrate out. Contact time of 30 to 40 minutes is suggested to obtain good results. This contact time can usually be obtained with flowrates of the regenerants in the range of 0.25 to 1 gpm/CF.

Regenerant Level. Regeneration levels for the WBA resins will range from 110% to 150% of the equivalent capacity used and will depend to some extent on the type of regenerant used and on the particular type of WBA resin (that is, as to whether the resin does have some strongly basic capacity or is all primary, secondary, or tertiary amine functionality). These regenerating levels may range from 1.5 to 2.0 lbs/CF for ammonia (NH_3), 3.0 to 4.0 lbs/CF for ammonium hydroxide (NH_4OH), 3 to 5 lbs/CF for caustic (NaOH), or 3 to 6 lbs/CF for sodium carbonate (Na_2CO_3), all expressed on a 100% chemical basis.

Capacity and Capacity Factors. The capacities that can be obtained from the WBA resins depend to a considerable degree upon the type of WBA resin as well as on the other factors influencing the capacities of weak electrolyte resins. (Weak electrolyte is another term used for weakly basic and weakly acidic; strongly acidic or strongly basic resins are referred to as strong electrolyte resins.) These other factors are: temperature of water being treated; flowrate during service cycle; and percentage of total anions that are sulfates, chlorides, and alkalinity (or free CO_2 after water has been treated by SAC resin).

As can be recognized from previous discussions on effect of temperature and flowrate (or contact time), the higher the temperature, the faster the exchange reaction will take place; and the longer the contact time, the less spread-out will be the exhausted band in the resin bed. The result will be greater capacity, because the slippage or leakage of anions (in this case chlorides) that would signal a conductivity or pH end point would not come as rapidly.

Sulfates are more strongly held by the resin than chlorides, and they can also load to a greater extent than would be calculated from the raw water analysis where sulfates would be expected to load as the divalent anion SO_4^{-2}. However, when the pH conditions on the exhausted portion of the resin bed become more acidic, sulfates can load as the monovalent bisulfate (HSO_4^-), almost doubling the capacity that would be obtained for the divalent sulfate ions. The higher the sulfates in the water being treated, the higher the total capacity will be.

The alkalinity, or alkalinity that has been converted to carbonic acid (or carbon dioxide in solution if the system is kept under pressure) by treatment with the hydrogen form of the SAC resin preceding the WBA resin, will also have an influence on the capacity that can be obtained by some types of WBA resins. These factors can be shown graphically as follows:

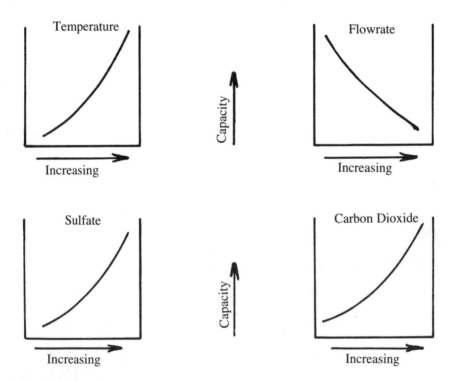

Figure 6-9. Effect of various factors on the capacity.

The effect of FMA or total strong anions on capacity varies with the type of WBA resin being considered. Generally the more weakly basic resins will show an increase in capacity as the FMA increases, but those resins with greater basic strength will show a capacity decrease as the FMA increases.

With the large number and types of WBA resins available, the many factors influencing the capacities that can be achieved, and the variety of regenerants that can be used, it is most difficult to state any typical capacities and efficiencies, other than to give some general ranges. Using caustic as the regenerant, capacities of 20 to 38 Kg/CF might be obtained at efficiency levels of 0.15 to 0.17 lbs/Kg.

Displacement (or Slow) Rinse
After the introduction of the regenerant chemical to the resin bed has been finished, there remains a resin-bed volume of regenerant in the resin unit that has to be displaced to finish the regeneration. This last resin-bed volume of regenerant is the best regenerant the resin will see since it has not been diluted by water left in the resin bed at the start of regeneration and does not have as much of the waste or contaminating ions in it. It is most important to fully utilize it by displacing it at the same flowrate as was used in the original regenerant introduction.

This is usually done by shutting off the flow of regenerant chemical and continuing the flow of the dilution water. In cases where a very concentrated brine or other regenerant has been used in the regeneration it will be necessary to shut off the regenerant chemical and start the flow of the rinse water to keep the same flowrate as that of the regenerant chemical through the displacement rinse.

Quality of Rinse Water. The displacement rinse water should be of the same quality as the dilution water used in the chemical regeneration. That is, if softened water were required as the dilution water for the chemical introduction, it should be continued through the displacement rinse.

There are some cases (such as cation resin regeneration in the hydrogen form) where raw water is used as the dilution water for the acid, but because of high hardness it would not be considered best to use this in the displacement rinse of the resin. This is because even at the lower acid concentrations, there are possibilities that calcium could be picked up from the rinse water in sufficient amount to reduce the capacity to a slight degree in the service cycle. In most cases this would not be a problem.

In the case of anion resin regeneration, displacement rinse, and final rinse, it is necessary to use softened, decationized, or deionized water to prevent calcium or magnesium precipitation in the high pH regenerant and regenerated anion resin environment. If a separate softener unit is not available to provide this water, then decationized water from the cation unit, either at the end of the service cycle or after regeneration and rinse of the cation unit, can be used for the anion resin regeneration and rinse. The added capacity of the cation resin required to produce this additional volume of decationized water in addition to the full service cycle must be achieved by increasing the resin volume and regenerant requirements of the cation resin unit. Where very high quality produced water is required, it may be desirable to use DI product water for regeneration and rinsing of the anion resins.

The displacement rinse need not be much longer than the volume of water to displace the void volume (spaces between the resin beads) of the resin bed, which is roughly 40% of the resin bed volume plus any additional tank volume from below the regenerant distributor to the top of the resin bed. This amount of rinse, however, would still leave behind some of the regenerant that has soaked into the resin beads, and would require even more slow rate rinse to leach out the remainder. This additional slow rinse, over that of one void volume, usually can be achieved in an additional one half to full void volume of rinse water.

In the case of the SBA resin, where hot dilution water has been used to improve the removal of silica during the regeneration step, the continuation of hot water through the displacement rinse is desirable. As has been stressed, the displacement rinse is just the continuation of the regeneration process.

Fast or Final Rinse

If time were of no importance in completing the regeneration process, and the distribution of rinse flow were very good, it would be of some advantage to continue rinsing the residual regenerant chemical out of the resin bed at a slow flowrate to improve the leaching of the regenerant from the resin beads. Time is of some importance, however, since the longer the resin units are out of service, the more or larger units would be needed to maintain service flowrates. Also, because of some imperfections of distribution or presence of some "dead" areas in the tanks, it is desirable to increase the flowrates for final rinse to flush out regenerants from these areas. The higher flowrates can give more turbulent flow in the "dead" areas such as in the areas under the bottom distributors, in the small radius curvature where the dished head meets the straight wall of the tank, where the internal piping or distributor supports obstructs flow, and so on.

The fast or final rinse is in many cases done at the flowrate used in the service cycle and with the feedwater that the particular resin unit would treat in the service cycle. For a softener system or lead cation unit in a DI system, this would be the raw or pretreated water. In the case of DI systems, the lead cation resin unit would be fast rinsed until a desired quality water out of that unit were reached, as measured by pH and/or conductivity or sodium leakage, and then the effluent from that unit would be used to rinse the anion unit next in line to its quality end point, and so on through whatever train of units might make up the total system.

Provisions for the added capacity needed to handle the additional water used for rinsing resins downstream must of course be provided. There are numerous possibilities and modifications which can be used, depending on the complexities of the systems and whether the units in a train are regenerated as a train or as individual units. A train or individual units may be put on line as soon as the quality end point of rinse is reached, or may be shut down and placed in standby until they are needed on-line.

The fast rinse flowrate is normally recommended at around 1 to 1½ gpm/CF if not done at the service flowrate of around 2 gpm/CF.

Recycle Rinse. With increasing demands to reduce volumes of waste water and since the fast or final rinse constitutes one of the major water uses for regeneration of ion exchange resins, there has been a growing use of recycle rinse. The recycle rinse can be employed where there is a cation unit followed by an anion unit.

The recycle rinse can be initiated after the conductivity in the rinse-down of the cation unit has been reduced to approximately 2,000 micromhos (500 ohms/cm^3 resistance) and the displacement rinse of the anion unit has been completed. Piping the effluent from the anion unit through a pump to the influent of the cation unit will allow recycling of the water from the cation unit to rinse down the anion unit, but instead of going to drain it will pass back through the cation unit. This excess caustic from the anion unit will be removed as the water passes through the cation unit, with the excess acid from the cation unit being removed by the anion unit, to eventually result in better quality water being used in each cycle of the water passage through the two beds, until DI water quality is reached.

In the process, some capacity of each bed is used up, but compared to the capacity used in continued rinse with raw water, where the ionic load or TDS remains constant, there will be less capacity lost (particularly if the raw water has a high TDS).

Volume of Rinse Water. The volume of rinse water used for each resin will vary considerably, depending on the regenerant level as well as on the resin structure and quality end point to be reached.

The general ranges are:

        25 to 75 gal/CF for SAC resins
        30 to 75 gal/CF for WAC resins
        50 to 75 gal/CF for SBA resins
        25 to 75 gal/CF for WBA resins

Long rinse-down times may be seen with WBA resins due to oxidation attack on the resin resulting in the formation of weak acid groups on the resin. These carboxylic groups on the resin will pick up sodium from the caustic regenerant, resulting in slow sloughage or hydrolysis of the sodium from these weak acid groups during the rinse to quality. Switching to ammonia as a regenerant in place of caustic will reduce this problem. There are some WBA resins that have been developed that are more resistant to this oxidation attack than others. Other influences on rinse requirements, particularly with anion resins, are fouling (both organic and physical), oxidation attack, and cross-contamination of anion with cation resin. There will be more discussion of these problems in Chapter 9 on troubleshooting.

CHAPTER 7

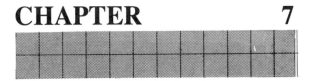

ION EXCHANGE SYSTEMS

Following the order of applications development in water treatment for the use of ion exchange resins, let us look at some of the ion exchange systems and their operations to give some idea of the many choices that may be made in systems designed to fit the end needs.

Water Softening
The first and probably still the largest volume use for ion exchange resins is for the softening of water. In the previous parts of this text the softening process has been used to illustrate ion exchange reactions and processes and has been discussed rather completely.

SAC resins are used for most water-softening applications. The WAC resins, however, are finding an increasing use where water conditions of high TDS cannot be properly treated with the SAC resins alone, or at all. Indeed, there are many ion exchange applications where it is desirable to use more than one resin to do the job. There are applications where other water-treating processes are aided or improved by ion exchange, and sometimes the reverse is true.

One example of the combination of ion exchange softening with other processes is the polishing or removal of residual hardness after cold or hot process softening, which are precipitation processes. Cold lime or lime-soda softening can be used to remove calcium or calcium and magnesium hardness, depending on the water composition, to a level of 35 to 80 ppm hardness, while hot lime-soda process can achieve a hardness level of 10 to 40 ppm. Ion exchange softening on either the cold or hot process effluent can readily reduce the residual hardness to less than 1 ppm.

An example of ion exchange softening as pretreatment to another process is the use of an ion exchange softener for removing hardness to reduce or eliminate scaling problems for reverse osmosis (RO) systems. This same system with RO may also include a DI system, following the RO, to remove the ions still remaining in the product water of the RO in order to give better economies than are possible using DI by itself.

Figure 7-1. Outdoor, dual-automatic demineralizers.
Photo courtesy L*A Water Treatment Corporation.

For water softening, as well as for most all other ion exchange applications, ion exchange resin manufacturers provide numerous manuals, data sheets, and brochures, which will allow the preliminary evaluation, design, and sizing of ion exchange systems to treat a given water to meet quality requirements. Most of the data compiled and published by the ion exchange resin manufacturers on operation of ion exchange resins has been obtained under carefully controlled laboratory conditions. Application of these data to commercial and industrial equipment may require some adjustment based on the equipment used and on practical experience. The engineering companies that manufacturing ion exchange units and systems can draw from their experience and engineering know-how to properly design equipment needed to do the job.

Development of New Operation Procedures

Up to this point we have been using a co-current basis for discussing ion exchange systems. Let us now examine other modifications of the co-current operation and countercurrent operations, once again using the softening process as the example. The main reason to look at other processes or operations is to improve the quality of the product water and to improve efficiencies. For softening, it would be desirable to reduce the hardness leakage.

To review the co-current operation (downflow service cycle and downflow regeneration): On completing the brine regeneration, the bottom portion of the resin bed is not regenerated as completely as the top portion of the resin bed. There will be a band of only slightly regenerated resin on the bottom of the resin bed. When softened water starts down the column in the subsequent service cycle, it will tend to remove a small amount of the hardness, which will show up as hardness leakage, from this bottom band. This can be illustrated by the following set of curves:

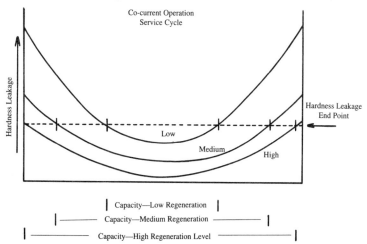

Figure 7-2. Co-current operation service cycle.

As can be seen, the leakage of hardness at the beginning of the run is due to the hardness remaining on the bottom of the resin bed with co-current regeneration. The hardness leakage at the end of the run is due to the leading edge of the hardness exhaustion band breaking through the bottom of the resin bed.

Prior to the popular usage of countercurrent regeneration systems, there was promoted a system that improved on co-current regeneration and approached the objectives of improved water quality by using an air mixing of the resin bed after regeneration to obtain lower leakage and improve regeneration efficiency (Unimixing by Rohm and Haas Co., 1954).

The air mixing is accomplished by introducing air into the bottom distributor at a pressure of about 2 to 3 pounds per square inch and a flow of 5 CF per square foot for about 15 minutes, Air mixing after regeneration results in breaking up the layer of poorly regenerated resin at the bottom of the resin bed and distributing it throughout the resin bed, and also brings some of the the well regenerated resin from the top of the bed to distribute it throughout the whole bed.

The result of this air-mixing regeneration approach on the quality of the water produced or leakage during the run could be shown for a water softening system with co-current regeneration as follows:

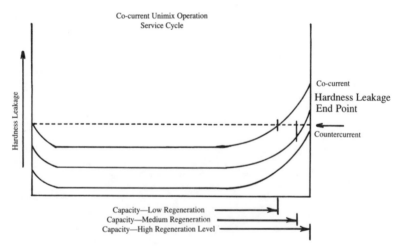

Figure 7-3. Cocurrent unimix operation service cycle.

This would occur because there would no longer be a band of hardness on the bottom of the bed to cause the hardness leakage at the start of the service cycle.

This air-mixing process also would be of benefit for SAC resins regenerated to the hydrogen form with sulfuric acid by mixing both before and after regeneration. Air mixing prior to regeneration allows the use of sulfuric acid at higher concentrations because the concentrated bands of calcium would be broken up and more evenly distributed throughout the entire resin bed. Thus the solubility limit of calcium sulfate would not be as easily exceeded and the result would therefore be higher capacity without calcium sulfate precipitation.

Countercurrent Operation
The advantages of countercurrent operation have been recognized since the early conception of ion exchange systems, but the mechanics of adopting it to operating resin systems has been somewhat slow in evolving. Presently it is becoming more popular, with the mechanical problems mainly overcome. The obvious advantages are lower operating costs, better quality of product water produced, lower wastes, and in some cases lower capital costs.

First, let us look at some of the advantages in countercurrent regenerating and again use water softening as an example. With the conventional downflow service cycle system, the top of the resin bed is almost totally in the calcium and magnesium form, with the bottom of the bed still partially in the sodium form. To regenerate countercurrent to the service flow would require introducing the brine regenerant upflow (from the bottom to the top). This would mean that the fresh brine would remove the small amount of hardness from the bottom of the bed very completely and still have good strength of brine to start removal of hardness as the regeneration proceeds up through the bed. At the end of the brine regeneration, the least regenerated section of the bed would be near the top, while the bottom of the resin bed would be very thoroughly regenerated.

The displacement rinse would have to be done using softened water to prevent hardness deposits on the bottom of the resin bed. Otherwise these deposits would show up as hardness leakage at the service cycle downflow. The final or fast rinse can be done with raw water downflow to a salt-free or hardness end point.

The hardness leakage during the start of the service cycle will be practically eliminated, with extremely low hardness leakage through the entire service cycle, since the very well regenerated resin at the bottom portion of the resin bed will remove most all traces of remaining hardness that may not have been removed in the upper portion of the resin bed. The hardness leakage at the end of the service cycle will occur very rapidly when it is reached, and there will be a greater utilization of capacity for the amount of salt used in regeneration than there is in co-current regeneration (if all conditions are met for the regeneration cycle to keep the bed compacted).

The quality of water produced by the countercurrent operation is difficult if not impossible to obtain with co-current operation and, to a much greater degree, is less dependent on the regeneration level. The regeneration level with countercurrent operation will affect the run or service cycle length, and also the quality of produced water, both to a considerable extent.

Comparing countercurrent operation to co-current (even with air mixing) operation, the service cycle would be shown as follows:

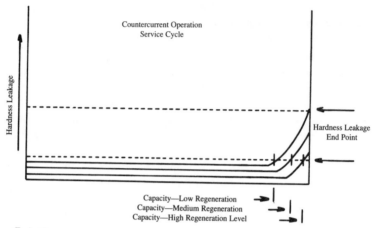

Figure 7-4. Countercurrent operation service cycle.

Another advantage of countercurrent (over co-current) regeneration in water softening operations is that even high TDS waters can be treated with lower hardness leakage. Countercurrent polishing units may therefore be used to good advantage in series with co-current regenerated units. In this type of system, the brine used for regenerating the polishing countercurrent unit may be used in series for regenerating the co-current unit to obtain even better salt efficiencies.

Usually it is found that the fast or final rinse to quality can be reduced with countercurrent regeneration, thus reducing the amount of waste water.

The key to proper operation of countercurrent systems is to be able to regenerate the resin in a compacted state. To regenerate in a countercurrent system (assuming downflow in the service cycle) the regeneration must be performed upflow. If the regenerant (in this case brine) is introduced upflow through the resin bed, there will be expansion of the resin bed (similar to what happens in backwashing); and because the regenerating solution is higher in specific gravity or density there will be more resin-bed expansion than would take place with water at the same flowrate. This expansion will cause the resin beads to separate from each other, with space or voids being formed through which the regenerant flow will not make as complete contact with the resin beads. The result is inefficient use of the regenerant.

To overcome this and to retain a compacted resin bed, some means of holding the bed immobile must be found to obtain the benefits of the countercurrent regeneration. A wide assortment of mechanical design innovations have been developed to achieve this goal, a few of which would include:

- A blocking flow of water that moves from the top of the unit to a takeoff distributor positioned at the top of the resin bed. This distributor removes the upflow regenerant solution. The blocking flow prevents the bed from expanding above the takeoff distributor.
- A buried takeoff distributor below the top surface of the resin bed.
- An inflatable bladder or bag which can restrict the movement of the resin bed, either by pushing the resin out and upward to fill the freeboard area or by holding down the resin from the top.
- Reversing of the service flow to go from bottom to top of the unit, instead of top to bottom, and at sufficient velocity to keep the resin compacted against a top distributor. Regeneration is then downward as the resin comes to rest and compacts against the bottom distributor.
- This same upflow service cycle can also be done by use of an orifice plate, where the resins will be supported by the velocity of upward flow, but the resins can drop through the holes in the plate when the upward flow is stopped for downflow regeneration.

It is important, for best results, to keep the resin bed classified, not disturbed, and in the same configuration as it was during the service cycle for the regeneration, as well as when going from the regeneration cycle back to the service cycle. This is to maintain the exhausted bands and the well regenerated sections of the bed positioned for maximum capacity and highest quality produced water in the countercurrent operation.

Very superior quality water can be produced by countercurrent operation as compared to co-current with leakage values in the ppb range for countercurrent as compared to ppm range for co-current. These advantages are usually more important for DI applications than for softening, and might eliminate polishing ion exchange units that are required to produce higher quality than co-current units are capable of producing.

The displacement rinse in countercurrent operation *must* be done with water that does not contain the ions that might reload on the zone of the resin bed which is most completely regenerated, or the leakage of these ions will appear in the early part of the service cycle. This means that for SAC resins in sodium form for softening applications, softened water should be used. In hydrogen form for deionization, decationized or DI water should be used. For

SBA resins it is best to use DI water, but reasonably good results can be obtained with softened water, since the pH is still quite high even after the displacement rinse, and silica would not load onto the resin under these conditions.

SAC Resins—Hydrogen Cycle

SAC resins in the hydrogen form are used principally in the lead unit in deionizing systems, but they also are used for the dealkalization of water for boiler feed, process water, and cooling-tower water applications. Waters with high alkalinity can be treated with the hydrogen form of SAC resins, which will replace the cations associated with alkalinity with hydrogen to form carbonic acid (H_2CO_3). Carbonic acid will convert to its constituents of water and carbon dioxide, and the carbon dioxide is released through forced-draft degassing. However, the chlorides and sulfates also present in the water would be converted to their corresponding acid forms (HCl and H_2SO_4), resulting in an acidic product water that would have to be neutralized with caustic to prevent having a corrosive water.

This neutralization could also be accomplished by blending of the decationized water with softened feedwater as shown in Figure 7-5.

The feedwater flow is split, with a portion going to the hydrogen-form resin unit while the other portion goes to the sodium-form resin unit, and then the effluents of both units are rejoined to produce near-neutral dealkalized water after decarbonation. This system is called "Split-Stream Dealkalization." It is of course necessary to calculate from the water analysis the correct split between the water flow going to the hydrogen-form resin and that going to the sodium-form resin so that a neutral blended effluent will result. For pH control there may be a small amount of $NaHCO_3$ allowed in the product water.

Dealkalization of water may also be achieved by the use of SBA resins operated in the chloride or chloride-hydroxide cycle. Higher capacities can be realized by use of type II rather than type I SBA resins. Although bicarbonates are not held as tightly as chlorides on the SBA resins in the hydroxide form, when the resin is predominantly in the chloride form and the pH has been raised by a small addition of caustic to the brine regenerant, there will be a favorable exchange of bicarbonate for the chloride. This exchange works well only with high alkalinity waters (40% to 80%), with capacities of 4 to 10 Kg/CF being obtained. The advantages of SBA resin dealkalization is that low-cost salt is used in place of the acid necessary for the SAC resins, and unlined steel tanks can be used.

WAC Resins

Since WAC resins in the hydrogen form do not have any great ability to split neutral salts, the majority of their capacity is for removal of cations up to the

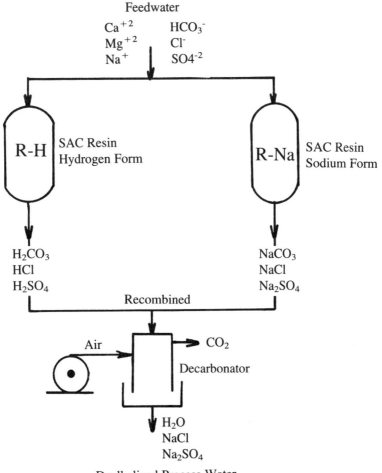

Figure 7-5. Split-stream dealkalization.

amount of alkalinity present in the water, and therefore they are excellent candidates for dealkalization. With the preference for divalent ions over monovalent sodium ions, the capacity is greatest (up to 60 Kg/CF) when the hardness is equal to or greater than the alkalinity of the water. Some of the WAC resins have good capacity for removing sodium alkalinity (up to 16 Kg/CF), which is much less than their capacity is for calcium and magnesium alkalinity.

The WAC resins are very flow- and temperature-sensitive, and their capacities are influenced by the TDS of the water. High flowrates, low temperature, and low TDS will result in lower capacities with WAC resins.

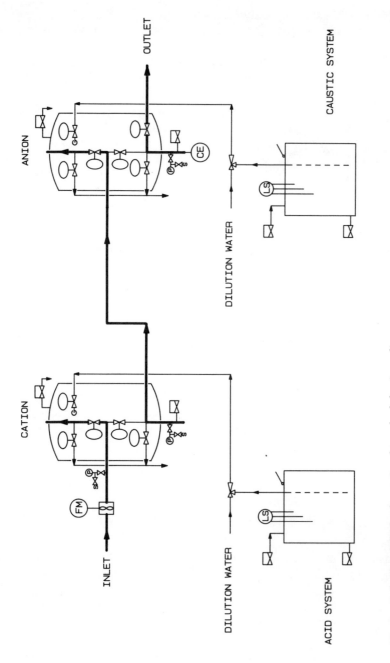

Figure 7-6. Simplified schematic of two-bed demineralizer.
Courtesy of L*A Water Treatment Corporation.

The advantage of WAC resins for dealkalization over SAC resins is the potential for one unit to do the job at higher capacities (depending on the water analysis).

With waters having alkalinity over 20% of the total anions, and a hardness-to-alkalinity ratio of one or greater, use of WAC resins in the hydrogen form preceding the SAC resin in the hydrogen form may provide higher total capacity as well as economy in regenerant usage, as compared to use of the SAC resins alone. This is particularly true if the excess acid from the regeneration of the SAC resin is used to regenerate the WAC resin. With sulfuric acid as the regenerant (because of calcium sulfate precipitation problems), this regeneration procedure could be used successfully if the SAC resin does not have any great amount of calcium loaded on it as a result of the hardness-to-alkalinity ratio being greater than one in the water being treated. With hydrochloric acid as the regenerant, there would not be any problem with hardness-to-alkalinity ratios above one.

Systems have been developed and used that install both WAC and SAC resins in a one-tank unit. These achieve the same results as when the WAC resin tank or unit precedes the SAC resin unit. Since the WAC resin has a lower density than the SAC resin, the WAC resin can be layered on the top of the SAC resin. However, the density differences are very small in the exhausted form of the two resins, so that it is desirable not to backwash the resins after exhaustion, or the layers of resin will mix. Since the bead size of the WAC resin changes from large to small when it goes from the exhausted to the regenerated form, the hydraulic separation can be maintained by backwashing only after regeneration. The best separation, regeneration efficiencies, and quality of produced water are maintained by regenerating and rinsing with decationized or DI water upflow.

SAC + WBA—Two-Bed DI

We have looked at the cation and anion resins separately, and now we will put them together in a DI system. With the cation resin exchanging hydrogen ions for the cations in the water to form the acids of the anions in the water, all that is necessary is for the WBA resin to remove the mineral acids formed, with the bicarbonates and silica passing on through.

Any leakage of sodium from the cation resin will combine with the chloride anion to form the neutral salt, which cannot be split by the WBA resin and which therefore will appear in the effluent along with the carbonic acid and silica. The pH of the effluent water after degassing would therefore be 7 or below. Degassing may not be required for many applications of this quality water.

SAC + SBA—Two-Bed DI

The main difference between this two-bed system and the preceding one is

Figure 7-7. Two-bed demineralizer followed by mixed-bed polisher.
Photo courtesy L*A Water Treatment Corporation.

that essentially all of the anions would be exchanged for the hydroxyl anion on the SBA resin, resulting in a product water containing only the sodium leakage from the SAC unit (as NaOH) and whatever silica leakage may occur. The effluent pH will be above 7 and would normally start out near 10 and drop to pH 8 for a majority of the run (depending on the water analysis, regeneration level, and sodium leakage from the SAC resin), with the end point of the run occurring at the point where there is a pH drop to below 7. At approximately this same point of pH drop there will usually occur a very slight increase in conductivity, followed by a sharp decrease in conductivity, and then immediately followed by a gradual increase in conductivity as the silica, the bicarbonate, and the chloride leakages occur.

To increase the run length and reduce the load of anions on the SBA resin, a degassing system may be used to remove the carbon dioxide resulting from the breakdown of carbonic acid in the low pH or acidic effluent from the SAC resin. The higher the percentage of alkalinity in the raw water, the more favorable are the economic conditions to use a degasifier. This reduces the amount of SBA resin needed to handle the ionic removal of the carbon dioxide and the added amount of caustic needed to regenerate the SBA resin. The degassing does require capital cost, power cost for either forced-draft fans or vacuum pumps, and pumping to provide the pressure to transfer the water to and through the SBA resin unit. Obviously there is a break-even point where it will be cost-effective to degas. This will depend on the amount of water to be treated and the percent alkalinity in the water.

The application of countercurrent regeneration to any of the SAC resin and SBA resin systems discussed in this section on ion exchange systems will of course improve the quality of product water that can be obtained. In fact, it could improve the quality to the point where it would be difficult to match even with a four-bed system (2 two-bed systems in series) using co-current operation. This regeneration technique should be kept in mind for upgrading the performance of the many of the systems discussed here.

SAC + WBA + SBA—Three-Bed DI

This system is the addition of a SBA resin to the previously discussed SAC + WBA resin system. This addition will upgrade the quality of the water produced by removing the carbon dioxide and silica leakage and by removing also the chloride that was the result of the sodium leakage from the SAC resin. Now only the sodium is left as sodium hydroxide in the effluent. For high alkalinity waters, the addition of a degassing device after the WBA unit would further reduce the anion load going to the SBA resin.

Another advantage of this three-bed system over the SAC + SBA two-bed system is the economies possible in using the excess caustic from the SBA resin regeneration to regenerate the WBA resin. A precaution must be observed in this series regeneration procedure: Since the SBA resin will only have the weak anion load of bicarbonate and silica and perhaps the small load of chloride that leaks with the sodium from the SAC resin, the major load may be silica. With a large amount of silica coming off the SBA resin during the early portion of the regeneration as sodium silicate (Na_2SiO_3), it could be precipitated in the WBA resin unit, because the pH of the WBA resin will be low or acidic after exhaustion with sulfuric and hydrochloric acids. In this acidic environment the silicic acid will polymerize and precipitate in or on the WBA resin.

There are two ways to get around this problem. The first is to direct to drain the first portion of the regenerant containing the largest amount of silica and to use roughly the last third of the waste caustic to regenerate the WBA resin. The second way to prevent this silica precipitation is to dilute the waste caustic to a concentration below 2%, which is still a strong enough concentration to regenerate the WBA resin, but a low enough concentration to prevent the silica from precipitating out of solution. Calculations should be made to ensure that there is sufficient excess caustic in the SBA resin regeneration to complete the regeneration of the WBA resin. If there is not, the caustic going to the SBA resin must be increased accordingly. This will improve the degree of regeneration of the SBA resin and will provide a satisfactory regeneration of the WBA resin.

As was pointed out for the SAC + SBA two-bed system (and the same will apply to the SAC + WBA + SBA system), the main leakage is the sodium that leaked from the SAC resin, which will be present as sodium hydroxide in the final product water. If the resulting high pH due to this leakage is objectionable, it can easily be reduced by the use of a polishing WAC resin in the hydrogen form—one with an extremely high capacity (about 60 Kg/CF) to remove the sodium.

A further modification of this three-unit system, which will reduce the number of tanks from three to two, is layering of the WBA resin on the top of the SBA resin in one tank, similar to the previously discussed system where a stratified or layered bed was used to place a WAC resin on top of a SAC resin in one tank.

This approach can work only if there is enough specific-gravity difference between the WBA resin and the SBA resin to keep the WBA resin on top, so that it will remove the free mineral acidity (FMA) and leave the weak anions to be removed by the SBA resin. Not all WBA resins can be used in this type of system as they do not all have this low specific gravity. Also, to maximize the separation it is desirable to screen the resins to be used to remove the fine particle size resin from the SBA resin on the bottom and to remove the very large particle size resin from the WBA resin on the top.

Countercurrent regeneration will be possible with this layered-bed approach. Greater economies on caustic use and excellent water quality can both be achieved. The potential for silica precipitation with countercurrent regeneration exists, as discussed in the separate-tank approach; and the same solution to the problem can be used.

Not all types of waters can be treated by this type of system, as there can be problems with using minimum bed depths of one or the other resin while maintaining balanced capacities within a reasonable total resin bed depth.

In the three-bed (or two-bed with layered anion unit) system, another potential advantage for having the WBA resin in front of the SBA resin is that it will help remove the organic foulants that may be present in the raw water and will prevent them from loading on the SBA resin, which is more susceptible to fouling than are the WBA resins.

As can be seen, there are a very large number of combinations of the four basic ion exchange resins that can be put to good use, depending mostly on the type of water to be treated, the end quality of water required, the economies of regenerant desired, and other special problems such as organics or other foulants, capital costs, and operating costs.

The flow requirements also will have an influence on the complexity of the system that could be afforded to achieve the end performance goals. Thus far we have discussed the basic systems and some possible modifications. Below are summarized some of the other multibed systems, with brief comments regarding any unusual characteristics.

WAC + SAC + SBA—Three-Bed DI

This system can include degassing preceding SBA, and is useful for waters containing a very high percentage of hardness alkalinity.

SAC + WBA + SAC + SBA—Four-Bed DI

This is a combination of 2 two-bed systems to obtain higher quality water. The product of the SAC + WBA has a small amount of sodium leakage as NaCl, and most of the weak anions are not removed, so there is little load going to the following SAC + SBA resin units. The small amount of sodium can be reduced to a very small level of leakage by the SAC resin, which will allow the SBA resin to remove the weak anions remaining with only a very small amount of silica leakage possible.

This system could also have a degassing device installed following the WBA resin if the alkalinity were a major percentage of the anion load. However, with the use of waste caustic from the SBA resin regeneration going to regenerate the WBA resin, and the waste acid from the second SAC regeneration going to the first SAC resin, the economies of regenerate chemicals would make the degasifier difficult to justify.

With the present countercurrent-regenerated two-bed systems producing water quality comparable to the four-bed system, and with less capital cost and comparable chemical efficiencies, it might be well to make a close comparison of the two systems. The four-bed system likely can only be justified to treat very high TDS waters.

Most of the multibed systems that have been used in the past are no longer in as great demand since the advent of countercurrent regeneration systems and the use of mixed-bed ion exchange units.

SAC/SBA—Mixed-Bed DI

It is the writer's opinion (one shared by many others) that the development of the mixed-bed ion exchange resin system was one of the most important contributions to the application of ion exchange technology since the development of the resins themselves. Without this development it would not have been possible for many of the industries which require large volumes of better than 18-megohm-cm resistivity DI water to have come into being or to have developed as rapidly as they have. The semiconductor business and condensate polishing for the electric utilities are excellent examples of large new industries that are critically involved with the maintenance of ionically clean water that could practically and economically be supplied only by ion exchange.

Although there are possible combinations of SAC/SBA, SAC/WBA, WAC/SBA and WAC/WBA mixed-bed systems, by far the majority of installed systems are SAC/SBA. There are applications and installed systems for the other combinations of resins, but for brevity we will only go into detailed discussion on the operation of the SAC/SBA combination.

The key to the operation of a system using both cation and anion resins mixed in one tank is the ability to separate the two resins after the service cycle so that they can be regenerated separately (in the same tank or externally). Fortunately, in most cases, the anion resin is of lower density than the cation resin so that by backwashing, the lighter anion resin will rise to the top of the resin column and the heavier cation resin will settle to the bottom. Ordering resins with some of the small cation resin beads and some of the larger anion resin beads removed will make the separation sharper and more clearly defined.

The backwash step is most important to get a clean separation of the anion resin from the cation resin, with a minimum of mixing of the two, since the cation acid regenerant (HCl or H_2SO_4) would convert the anion resin in the cation zone to the completely exhausted acid form. The same problem would also be found with cation resin in the anion resin zone, where the sodium hydroxide regenerant for the anion resin would convert the cation resin completely to the sodium form.

Not only is this cross-contamination of the resins undesirable from the obvious loss of capacity resulting, but it can also cause reduced quality in the produced water. This impairment of final water quality results from the hydrolysis of the sodium from the contaminated cation resin that remains in the bottom of the mixed resin bed as well as hydrolysis of the chloride and sulfate from the contaminated anion resin that remains in the lower section of the mixed resin bed.

When the mixed-bed resin system is regenerated in the same tank used for the service cycle, there is an interface distributor located at the point where the cation resin zone interfaces with the anion resin zone after backwash and settling, as illustrated in the drawing below:

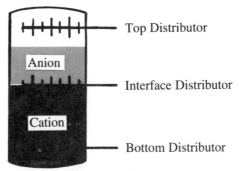

Figure 7-8. Cross section of mixed-bed column.

The flow of regenerants can be carried out in a variety of ways to achieve regeneration of the anion and cation resins with a minimum of cross-contamination of one resin by the regenerant of the other. Two such ways are illustrated below:

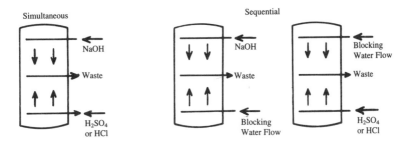

Figure 7-9. Flow of regenerants in a mixed bed-column.

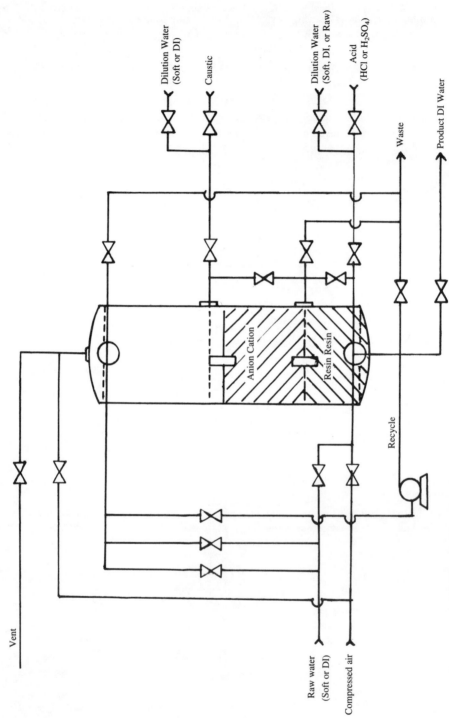

Figure 7-10. Simplified schematic of mixed-bed demineralizer.

It can be seen in the illustration below that there is bound to be some cross-contamination at the interface distributor:

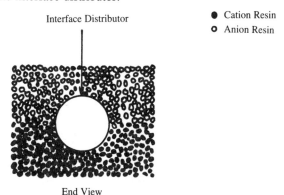

Figure 7-11. Cross-contamination at interface distributor.

Cross-contamination at the interface distributor can be the cause of approximately 10% loss in capacity, in addition to reducing product water quality.

One of the ways to reduce the cross-contamination occurring at the interface distributor is by use of an inert resin that has a specific gravity intermediate between those of the anion and cation resins and that will hydraulically classify between the cation and anion resins on backwash. With the proper quantity of this inert resin (approximately 10% to 15% of total resin volume) in the mixed bed, it will cover the interface distributor and effectively reduce the cross-contamination problem.

Another problem of operation in mixed-bed resin systems is to obtain a good mixing of the cation and anion resins after the regeneration and rinses are complete. The basic reason for using a mixed-bed system to produce high quality product water is that the intimate mixture of cation and anion resins allows for almost simultaneous removal of the cations and anions in the water being treated. Also, if there were some cation resin layer on the bottom of the mixed bed, the very small amount of sodium left on the resin (due to less than 100% regeneration) would tend to hydrolyze off the resin to give higher conductivity in the product water. A similar occurrence could also take place with the anion resin portion of the mixed bed, with silica or chloride (and even sulfate) showing leakage.

If a comparison is made of a mixed-bed system to a two-bed co-current system where the cation resins are separated from the anion resins in two tanks operated in series, it is immediately recognized that the produced water quality would be much poorer with the two-bed resin system. Now if we go to 2 two-bed resin systems in series (or a four-bed system), it can be seen that this latter system would produce much improved water quality. Carrying this addition of more two-bed systems in series to its extreme, we would end up

Figure 7-12. Dual-automatic, packaged mixed-bed demineralizer.
Photo courtesy L*A Water Treatment Corporation.

with produced water quality approaching that of the mixed-bed system.

However, although the mixed bed is similar to an infinite series of two-bed systems, it is dissimilar in that with a cation resin bead next to an anion resin bead, both the cation in solution and the anion in solution are removed almost at the same time. Thus there is a stronger driving force to shift the ion exchange reaction to completion by removing the competing ion from solution at the same time as shown below:

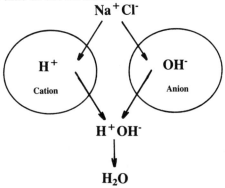

Figure 7-13. Mechanism for ion exchange in a mixed bed.

Mixing of the resins can be achieved with the help of compressed air or an inert gas such as nitrogen. Normal environment air contains relatively small amounts of carbon dioxide so the effect on the capacity of the anion resin in the air mixing step would be insignificant. Most problems with compressed air come from contamination by oil from the compressors. Oil contamination of the resins, even in small concentrations, coats the resin bead surfaces, which results in resin flotation or physical blocking of the surfaces and reduces ion mobility to the exchange sites.

Air or nitrogen gas mixing of the resins usually will require approximately 15 to 20 minutes at a flow of 7 to 10 scfm (standard cubic feet per minute) per square foot of bed area, depending on the configuration of the vessel and resin bed.

In the application of mixed-bed resin systems to the electric utilities for condensate polishing, it is common practice to transfer the mixed resins from the service vessels or tanks for separation and regeneration. This reduces the downtime (and therefore the number of service tanks required) for regeneration and minimizes the possibility of regenerant chemicals getting into the condensate system. The resins are usually remixed in external regeneration tanks before being transferred back to the service vessels.

An additional concern in this high flowrate service application (usually at 50 gpm/ft^2) is the filtering capabilities of the mixed-bed resin system for iron and other suspended metal oxides. The filtering capability of the mixed-bed system in condensate polishing is as important as its ability to remove ionic contamination, since both interfere with modern high-pressure utility generator operation. Frequently the termination of the service cycle will be based on pressure drop limits rather than on an ionic leakage end point.

To remove the particulate material from the resin bed, a very vigorous backwash conditioning of the resins is performed using both air scrub and water prior to the chemical regeneration. The physical strength of the resins is of greater concern in this application due to the physical abuse in the transfer, backwash, and service conditions than in most other mixed-bed applications. The field of utility condensate polishing is one requiring extreme care, and the purity of product is measured in limits of one part per billion of sodium, chloride, and sulfate.

Other Systems

In water treatment, as well as for hydrometallurgy, sugar processing, chemical process, catalysis, waste treatment, and other applications, adaptations by numerous modifications of the equipment discussed thus far have been proposed and used in attempts to improve the operations of the systems or quality of produced product, or to reduce costs.

There have also been developed systems that give continuous production of treated water (or other products as indicated above), and those which use special regeneration and operating techniques to enable waters of high TDS to be treated economically. It is not practical to list all of these, and doubtless there are still more being invented or proposed. We will mention only a few which have been utilized in full-scale operation.

As has been pointed out, countercurrent operation has advantages in efficiency and quality of water produced. In the "continuous" ion exchange equipment designs, this principle is utilized with the resin being pulsed in a direction counter to the flow of the water being treated. The regenerated resin first contacts water that has been previously contacted with resin, and is therefore mostly deionized. This step removes the last traces of remaining ions from the water. In the next zone, or section, to which the slug of resin is pulsed, it contacts water that has the ions partially removed. Finally, in the various zones down the column, the resin slug, which is now more fully exhausted, will see the untreated water.

The resin slug is then removed from the loading or service section of the column and is transferred to the regeneration column. There it contacts regenerant chemical (in a countercurrent flow) that has been nearly fully utilized, and in subsequent pulses moves down the column contacting only partially used regenerant, and finally fresh regenerant to drive the resin to a more fully regenerated condition.

Finally, in the rinse section, the slug of resin is countercurrently rinsed by moving or pulsing through that section of the column until it is ready to start

Chapter 7

the route all over at the loading or service section.

Variations in design and operation of these continuous-process systems are considerable, and the above description is not intended to describe them completely. The advantages of these systems are that they obtain high efficiency of regenerant usage, are able to handle high TDS, and maintain excellent quality product with minimum quantity of resin inventory for the flowrate.

The disadvantages of continuous-process systems are the complexities of the valving and instrumentation necessary to maintain optimum operation and the higher-than-normal physical abuse of the resins as compared to fixed-bed systems. There are a considerable number of these types of systems in use.

Some examples of such systems are:

Himsley Engineering, Ltd. Toronto, Canada
    Himsley Column

Japan Organo Co., Ltd., Tokyo, Japan
    Organo-Asahi Continuous Ion Exchange Process

National Institute of Metallurgy, Johannesburg, South Africa
    NIMCIX Column

Robert Porter Associates, Salt Lake City, Utah
    Porter Columns

US Bureau of Mines, Salt Lake City, Utah
    US Bureau of Mines Column

CHAPTER 8

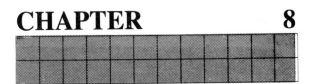

DESIGN AND SIZING OF ION EXCHANGE SYSTEMS

One of the ingredients of ion exchange water treatment has not been discussed yet: How to size or design a system or systems to treat a specific water so that the criteria of capacity and quality of produced water are met. To cover that in any complete fashion in this text is not possible. Some ideas and methods are given, however, so that the reader may be able to better understand how the designs and ratings of water treatment equipment manufacturers have been derived.

Ion exchange resin manufacturers have all done excellent jobs, for the resins they manufacture, of putting out detailed data on the physical, hydraulic, and operating characteristics for a wide range of operating conditions and water quality. In most cases, instructions are given on how to use these data to arrive at performance ratings for capacity and produced water quality.

Unfortunately, it is not always very clear under what conditions the data have been obtained, and there do not appear to be any standard reference conditions covering all of the resin manufacturers so that comparisons can be readily made. Also, the grades of resins in each category often are not identical. There are some few grades and types of resins that are produced only by one resin manufacturer, making the choice of resins still more complicated.

One thing common to all resin manufacturers is that practically all data has been obtained under laboratory conditions and in laboratory equipment. Practical performance in large-size systems is available only from experience obtained on equipment in the field and usually is available from the ion exchange equipment manufacturers who produce, install, and service the equipment. All ion exchange resin manufacturers readily admit to this difference between laboratory data and "real life" field performance, and offer some percentage correction factor or "fudge factor" that should be considered (usually from 80% to 95% of the laboratory data). For further confusion, laboratory data may be based on best performance, average performance, minimum specification performance, or performance with a built-in correction factor. The efficiency of the mechanical equipment in which the resin is installed will also have an important influence on the performance of the resins.

In summation of the above, experience with performance in full-scale field equipment is the best source of data for design or ratings of ion exchange systems performance. It should be pointed out, however, that performance of large ion exchange systems follows the data obtained in laboratory evaluations more closely than do most other chemical process systems. Close agreement has been obtained between performance using data collected from one-inch diameter columns and performance of large plant-size systems. Conservative design or sizing of ion exchange equipment or systems is desirable and in the long run may be very cost-effective and require less maintenance.

Softener Systems

Some preliminary discussions of softener design and sizing were covered in Chapter 5. To use a constant base for these discussions, let us use the water analysis that was used in Chapter 3 as an example for how water analyses can be expressed, and run through some sample calculations for an ion exchange system capable of producing 100 gpm continuously on a 24-hour-per-day basis with a maximum hardness leakage of 1.0 ppm.

Water Analysis in ppm as $CaCO_3$

Ca	100	HCO_3	82
Mg	103	SO_4	52
Na	130	Cl	199
			333
		SiO_2	10
	333		343

The total hardness is the sum of the calcium and the magnesium, or 100 + 103 = 203 ppm. We will want to convert this number in ppm to grains/gallon as $CaCO_3$, since the capacity of the ion exchange resin is given in kilograins as $CaCO_3$ per cubic foot of ion exchange resin. Since 17.1 ppm = 1 gr/gal, then 203 ppm/17.1 = 11.87 gr/gal. If we wish a continuous flow of 100 gpm, then 100 gpm x 11.87 gr/gal = 1,187 gr/min. This will be the hardness load to be treated.

Now let us look at the water quality that is required (1.0 ppm maximum leakage) and determine if this is obtainable and at what regeneration level for co-current operation. The sodium-to-hardness ratio of the above water is 130/203 = 0.64, and the TDS is 333 ppm. The sodium-to-hardness ratio in one resin manufacturer's data does not give values at 0.64 for hardness leakage, but by extrapolating the data we come up with the required regeneration level of 14 lbs/CF to give a hardness leakage under 1 ppm at a TDS of 333 ppm for a gel-type SAC resin of 8% cross-linkage. The capacity obtained at this regeneration level is shown to be 30.2 Kg/CF.

Using another resin manufacturer's data, the leakage of 1 ppm could be obtained at 10 lbs/CF of salt and a capacity of 31.2 Kg/CF. A third resin manufacturer's data show that the leakage of 1 ppm could be obtained with 10 lbs/CF salt level with a capacity of 26.4 Kg/CF.

At first glance, it would appear that there is a rather wide range of difference between resins of different manufacturers, and yet there are probably fewer differences in performance between resins of different manufacturers of this type of resin than there might be between individual batches from one manufacturer. With careful review of all the literature available and examination of how the laboratory data were obtained, the differences can be almost completely eliminated.

The effect of regeneration flowrate used, the end point used to determine run length or capacity, the service flowrate used, the bed depth used, the effect of the ratio of calcium to magnesium, the effect of the ratio of hardness to sodium, whether the leakage is an average or maximum allowable figure, the effect of TDS, and of course if there were any built-in safety factors in the published data: these are only some of the reasons for apparent discrepancies.

Using all the data available, we arrive at a regeneration level of 12 lbs/CF to obtain a leakage of under 1.0 ppm with a conservative capacity rating of 26.0 Kg/CF. That has also been adjusted to include a factor for converting laboratory data to full-scale industrial system operation. The data from all the resin manufacturers will all come out very close to the same values when the variables on how their data were derived are considered.

(The point to be made is that one should be sure all variables are known and how the data were derived before applying them to the calculations of an ion exchange system. The experience of equipment manufacturers with similar equipment and resin performance in the field is the best place to obtain reliable "real life" design and sizing information, since the equipment in which the resins are contained is vital to how well they perform.)

Now that we have established the capacity to use in this example for softening, and have a figure for the hardness load, we can figure the gallons of water that could be produced from a CF of resin by dividing the capacity per CF of resin by the hardness per gallon in grains per gallon or:

$$\frac{26,000 \text{ gr/CF}}{11.9 \text{ gr/gal}} = 2,185 \text{ gal/CF}$$

If we wished a minimum of 8 hours service run, then:

8 hrs × 60 min/hr × 100 gpm = 48,000 gallons should be treated per 8-hour service run. Then 48,000 gal/2,185 gal/CF = 22 CF of resin that would be required to meet this service cycle.

To test whether this quantity of resin would meet normal hydraulic requirements we will use a minimum bed depth of 30 inches or 2.5 feet. 22CF/2.5 ft = 8.8 sq ft area of the resin bed. Then 100 gpm/8.8 sq ft = 11.8 gpm/sq ft linear flowrate. This is a bit on the high side of desirable linear flowrate of 4 to 10 gpm/ft^2, but could be handled if there were no great amount of suspended solids or dirt in the water supply.

The space velocity or gpm/CF would be 100 gpm/22 CF = 4.5 gpm/CF, within the range of 2 to 5 gpm/CF considered desirable. The only problem that might be of concern in this configuration of resin bed would be the hardness leakage with a minimum bed depth of 30 inches with the linear flow rated at the high limits of desirability. We would therefore be inclined to go for a unit that would have a service run of 12 hours rather than 8 hours. This would require 12 × 60 × 100 = 72,000 gallons per service run. Furthermore, 72,000 gal ÷ 2,185 gal/CF = 33.0 CF of resin.

However, another practical consideration would be an odd-size tank. If we went with a 3-foot bed depth, that would result in a 11.0 ft^2 area or a tank with a 3-foot 9-inch diameter. With 12.5 ft^2 area, a standard 4-foot-diameter tank could be used, which 38 CF of resin would provide for a linear flow rate of 8 gpm/ft^2 and a space velocity of 3.0 gpm/CF, both well within normal design limits. The added CF of resin would actually give a 15% capacity increase, but could be compensated for by a slight reduction in regeneration level to 11 lbs/CF and still meet the leakage requirements.

For continuous service, this would mean a two-tank system with one on line and one in regeneration or standby for a total of 76 CF of resin. Regeneration time would consist of 10 to 15 minutes of backwash, 30 minutes of brining, 12 to 15 minutes of displacement rinse, and about 15 minutes of fast rinse, for a total maximum time of about 90 minutes.

An alternative to the above system would be to use three smaller tanks of 3-foot diameter, with two units on line in parallel and one unit off line for regeneration or standby. With 21 CF per tank, which would give 42 CF on line and therefore satisfy the linear flowrate and space velocity criteria, the total CF installed would be 83 CF, in place of 78 CF in the two-tank system. There would however be the cost of three tanks and the extra piping and valves. With changing flowrates required to service, or fluctuating supply water quality, the three-tank system might have advantages, as it would be possible to have only one tank on line or as many as three on line for some periods of time, if required.

SAC + SBA DI Systems

Cation Resin—Hydrogen Form. As compared to the softening or sodium-form design and sizing, there are some similarities, but there are many more variables that have to be considered in rating the capacities as well as the quality of produced water that can be achieved. For simplicity we will use the same water analysis, flow, and operating conditions as used for the softener systems. The calculations will be based on a system to produce a product water quality with 3 ppm average sodium leakage, average silica leakage 0.02 ppm, and 20 micromhos conductivity. Sulfuric acid will be used for the cation resin regeneration and sodium hydroxide for the anion regeneration. The same observation applies to variations which exist in data published by resin manufacturers for the resins described in the softener systems discussion. For the anion part of the DI system, the porous gel-type I SBA resin will be used.

The information needed from the water analysis for rating the capacity of the cation resin includes:

- Sodium concentration as percent of total cations: 39%
- Alkalinity as percent of total anions: 25%
- Calcium as percent of total cations: 30%
- Magnesium as percent of total cations: 31%
- TDS: 333 ppm

With the resin manufacturer's plotted curves of data and and the above information, we find that a regenerant level of 8 to 9 lbs/CF will satisfy the leakage requirements for sodium in all but one set of data. This one set of data shows a greater regeneration level (11.5 lbs/CF) to achieve the leakage level for sodium of 3 ppm. The possible explanation of this discrepancy is that a greater leakage end point was used in collecting their data, also resulting in a higher capacity. Additionally, higher flowrates were used. The other data is within a fairly close range with capacities of 18.0 to 22.9 Kg/CF. When all parameters are considered (including factors for full-size industrial equipment), we would rate the capacity at 16.6 Kg/CF at a regeneration level of 8.5 lbs/CF.

With 16.6 Kg/CF capacity and a cation load of 333 ppm or 19.5 gr/gal, we can obtain a number for gal/CF by 16,600 gr/CF ÷ 19.5 gr/gal = 851 gal/CF. However, before going further in our calculations, several adjustments should be made. One is the additional capacity that has to be used by the amount of rinse water (raw water) that is used in the rinse of the regenerant from the cation resin bed (roughly figured at 50 gal/CF). The other adjustment is for any decationized water that might be needed for the backwash, regeneration, and rinsing of the anion resin. (Very often, in rating two-bed systems, the engineer may start with calculations on the sizing of the anion unit first and then go to sizing of the cation unit so that the number of CF of anion resin is already known for the calculation of the added capacity of cation resin required for rinse of the anion unit.) If soft water is available, this can be used at less cost than the decationized water. We must also look at the amount of sodium leakage that will occur during the service cycle, since in order to be completely accurate, this amount should be deducted from the total cation concentration in the water analysis to show the actual load or capacity utilized by the resin.

In this case, 333 ppm of cations in the water analysis minus the 3 ppm average leakage will give 330 ppm actual cation load being utilized by the cation resin. If we then utilize the following formula, we can figure the operating capacity in gallons per CF of resin.

$$\text{Operating Capacity gal/CF} = \frac{\text{Rated Cap. - Cap. lost in rinse, etc.}}{\text{Total cations - Leakage}}$$

Or in this case:

$$\frac{16{,}600 - (50 \times 19.5)}{(333 - 3) \div 17.1} = 810 \text{ gal/CF}$$

Then 100 gpm × 60 min/hr × 8-hr run = 48,000 gal for 3 hrs.

The number of CF of cation resin would be 48,000 gal ÷ 810 gal/CF = 59 CF.

For a 3-foot bed depth, $59 \div 3 = 20$ ft^2 of area is needed (approximately). A tank with a 5-foot diameter has an area of 19.5 ft^2, so this would be adequate. The linear flowrate would then be 100 gpm/19.5 ft^2 = 5.1 gpm/ft^2, and the space velocity would be 100 gpm ÷ 59 CF = 1.7 gpm/CF, both well within limits. Before we can calculate the additional CF of cation resin to give a sufficient supply of water for the anion regeneration and rinse, we will have to calculate the anion resin requirements.

Anion Resin—Hydroxyl Form. The variables in the operation of SBA resins are to some degree greater than with cation resins, in view of the role that sodium leakage in the water coming from the cation resin has on the silica leakage from the anion resin, as well as the regeneration variables.

With the SBA resin, the primary control of quality is the silica leakage. In the example case, the water analysis shows 10 ppm as $CaCO_3$ with a total anion load of 343 ppm, or a silica percentage of total anions of 3%. The other factors that will be of concern from the water analysis are:

- FMA or TMA (the sum of chlorides and sulfates): in this case 251 ppm or 73%
- Alkalinity percentage of total anions: 24%
- Chloride ratio to FMA: 79%
- Chloride ratio to total anions: 58%

It is also necessary to decide whether optimum regeneration will be used (resin bed, diluted caustic, and displacement rinse each at 120°F). Other choices are dilute caustic only at 120°F, dilute caustic at 95°F, or dilute caustic at ambient temperature (75°F).

In this example we will go to 75 to 95°F regeneration, since the silica level is low compared to the TDS or total anions. Although the average silica leakage might be met at 75°F regeneration, the use of heated caustic (to prevent solidification) at 75°F with normal temperature water for dilution would bring the diluted caustic above 75°F, (in fact near 95°F), due to the heat of dilution of caustic.

Under the above conditions of water analysis and using all resin manufacturer's data available, we might rate the anion capacity at 10.4 Kg/CF at a regeneration level of 6 lbs/CF, which would include a safety factor for converting the laboratory data to full-scale equipment.

With the anion load of 343 ppm converted to gr/gal, $343/17.1 = 20.1$ gr/gal, the gal/CF would be $10,400/20.1 = 517$ gal/CF. For an 8-hour service cycle, to match the cation service cycle, 48,000 gallons need to be treated. Therefore 48,000 gal/517 CF = 93 CF of SBA resin are required. The 5-foot-diameter tank that was found satisfactory for the cation resin could also be used for the anion resin, with a 4.8-foot bed depth giving a linear flowrate of 5.1 gpm/ft^2 and a space velocity of 1.1 gpm/CF.

Now that we have the CF of anion resin required, we can figure the added capacity needed from the cation resin to produce sufficient decationized water for the anion regeneration, if soft water is not available. An amount of 50 gal/CF should be adequate for rinse requirements, and an additional 16.6 gal/CF would be required for regeneration dilution water for the caustic. This latter quantity is figured by calculating the total pounds of 4% caustic used in the regeneration per CF of resin, which would be 6 lbs $\div 0.04 = $ 150 lbs of dilute regenerant. Since this 150 lbs of dilute caustic contains roughly 6 lbs of 100% caustic, then $150 - 6 = 144$ lbs water are present. If 50% caustic is used to make up the 4% solution, then roughly the amount of water furnished with the 50% caustic would account for 6 lbs of water, so the dilution water required would be $144 - 6 = 138$ lbs, or 138 lbs/8.3 lbs/gal = 16.6 gallons of water per CF.

The displacement rinse of the anion resin at two bed volumes would be 7.5 gal/CF × 2 = 15 gal/CF, so 50 + 16.6 + 15 = 81.6 gal/CF. With 93 CF of resin times 81.6 gal/CF there would be 7,589 gallons of decationized water per regeneration. With the cation resin capable of producing 810 gal/CF, 7,589 gal/810 gal/CF = 9 CF of additional cation resin would be required. Now the cation unit would have 59 + 9 = 68 CF of resin in a 5-foot-diameter tank that would give a bed depth of 68 ÷ 19.5 = 3.5 ft.

If softened water were used for the SBA resin regeneration in place of the decationized water from the SAC resin, there would be a savings of 9 CF of SAC resin plus 76.5 pounds of sulfuric acid per regeneration. This would however require a softener tank and resin, plus salt for regeneration.

To summarize this two-bed system for the example water analysis:

- For 68 CF cation resin in a 5-foot-diameter tank with approximately 3.5 feet of bed depth, there should be 40% to 50% freeboard for backwash, so a 5-foot straight side on the tank should be adequate. The cation resin tank would be 60 inches by 60 inches, assuming no supports or underbed were used.

- The cation resin would be regenerated with 8.5 lbs of sulfuric acid per CF. Since the calcium concentration is 80% of the total cations, and the regenerant level is 8.5 lbs/CF, there would be a problem with calcium sulfate precipitation unless stepwise regeneration were used. It would be recommended that a two- or better yet a three-step regeneration be used with 2 lbs/CF of acid applied at 2% concentration, 3 lbs/CF at 4%, and 3.5 lbs/CF at 6%. The flowrate used for the diluted regenerant would be approximately 0.6 gpm/CF of resin for a total acid introduction time of 45 minutes. A total of 579 lbs of 100% acid (or 602 lbs of 66° Baumé acid) would be required per regeneration.

- 93 CF of anion resin in a 5-foot-diameter tank with an approximate 4.8-foot bed depth should have an approximate 70% to 100% freeboard to accommodate the backwash, so it would appear that a 9-foot straight side would be adequate. The anion resin tank would then have dimensions of 60 inches by 108 inches.

- The anion resin would be regenerated with 6 lbs/CF of 100% caustic. This caustic would be at a concentration of 4%, with the dilution water supplied by the regenerated cation unit (or softened water). At a regenerant flowrate of 0.25 gpm, this would require 66 minutes. The total caustic used would be 559 lbs of 100% caustic, or 87.7 gallons of 50% caustic.

- The regeneration time for this system would be:

	Cation (minutes)	Anion (minutes)
Backwash	10 to 15	10 to 15
Regeneration	45	66
Displacement rinse	19 to 25	44 to 60
Fast rinse	33	(33)
Dilution water for anion taken at end of service cycle at service flowrate	11	

Maximum time 209 minutes, or 3 hours and 29 minutes.

With an eight-hour service cycle, there will be no problem operating with a two-train system to obtain continuous service flow. By using a decationized water storage tank (or using soft water for the anion generation) and pump, it would be possible to cut the regeneration time by about 35 minutes, as both the cation and anion generation could take place simultaneously.

The above example is only a brief discussion of how ion exchange systems can be rated and sized for a particular water and quality of produced water. As previously mentioned, the variables in design are almost endless: countercurrent regeneration, degassing between the cation and anion units, WAC or WBA resins, and on and on ad nauseam. With experience, however, the choices can be narrowed down because the constraints of capital costs, chemical costs, quality of finished water, flow requirements, degree of automation, waste control, and other factors eliminate some of the available alternatives.

The wide variety of ion exchange resins available from an increasing number of resin suppliers (both foreign and domestic), with many of them not completely comparable to others in properties, makes more difficult the task of resin selection and adaptation to an ion exchange system. Understanding some of the principles of design and sizing of ion exchange systems will help with the tasks of selecting systems best suited to a particular need and choosing experienced engineering or manufacturing firms to do the detailed production of the systems. Also, a knowledge of the variables involved in the design and sizing of ion exchange systems will provide guidance in handling operations and performance problems that may occur.

CHAPTER 9

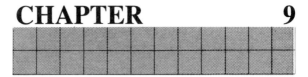

TROUBLESHOOTING

One of the conclusions that might be made, after reviewing the information and discussions presented thus far on the subject of ion exchange water treatment, is that the number of variables involved are almost astronomical. It is therefore not surprising that numerous things can go wrong, and usually do, at some time or other in the operation of ion exchange systems. Most of the time it is not just one thing, but a number of interrelated occurrences that cause the trouble. In discussions thus far, we have reviewed many of these variables and their effect on performance. In this chapter, a look at the sources of some ion exchange system problems may be an aid in troubleshooting.

Water

The infinite varieties of water and its impurities or contaminants are the source of a very large number of problems for ion exchange treatment. Foremost is the general fact that there are few water supplies which are constant in their composition from day to day or particularly over longer time periods. Designing the ion exchange system for the worst conditions of water supply is, of course, the safest approach. This will not eliminate all possible problems, however. New contaminants may appear, or there may be unexpected upsets in the water supply. Here are some of the problem-causers that can arise from the water source:

- Suspended solids and colloidal contamination, including clays, organometallic compounds, silica, iron, calcium sulfate, magnesium hydroxide, and aluminum hydroxide.

- Change in concentration and ionic balances or ratios of ionic contaminants such as sodium, silica, alkalinity, chloride, hardness, TDS, nitrates, manganese, iron, aluminum, phosphate, sulfide, and fluoride.

- Dissolved gases such as CO_2, H_2S, and SO_2.

- Organics such as vegetable decay products of humic and fulvic acids, lignins, proteins, organometallic compounds, esters, aldehydes, ketones, solvents, insecticides, bacteria, algae, and synthetic detergents.
- Oxidants such as chlorine, oxygen, chlorates, ozone, chromates, and others.
- Oil or greases, naphthanates, and other similar materials.
- Polyelectrolytes, emulsion breakers, scale inhibitors, filming amines, and sequestering compounds.
- Industrial and agricultural pollutants.
- Temperature changes, which can affect kinetics. (The rate of reaction doubles for every temperature rise of 10°C or 18°F.) Temperature changes also affect hydraulics, gas solubility, and thermal shock on resins.

Ion Exchange Resins

The great variety of ion exchange resins continues to increase. Most of them exist because they are better than their predecessors at solving problems of performance. The basic categories of course are SAC, WAC, SBA, and WBA resins. Within these resin groups, there are varieties of backbone structures: phenol-formaldehyde, styrene-divinyl benzene, acrylic-divinyl benzene, methacrylic-divinyl benzene, and many more. There are gellular and macroporous resins, varying degrees of cross-linkage, a variety of reactive groups, and other resin differences.

With such a variety of problems and resins, tracing the cause of trouble becomes most difficult.

Ion exchange resin life expectancy is a subject that deserves some discussion in view of its effect on the capital and operating costs of the systems. With the almost endless number of combinations of variables in the waters to be treated, the operating conditions, and the choices of ion exchange resins, it is possible to give only broad ranges of resin life expectancies. Life of all ion exchange resins is dependent on time, water flow, cycles of operation, exposure to physical abuse, chemical attack, temperature, foulants, operator training or performance, and equipment maintenance. Extremes of one or more of these influences can considerably shorten the usable life of the resins.

Cation exchange resins are generally based on the same resin structure as many of the anion exchange resins, but since they have more chemically durable exchange sites than the anion exchange resins, they also exhibit longer usable life. Under reasonable conditions, cation exchange resins may have a life expectancy of 5 to 10 years. Under extreme conditions, however, cation exchange resin life can be as little as 1 year and as long as 20 years.

Anion exchange resins are more susceptible to oxidation attack, high temperature, and foulants than cation exchange resins, and so have a life expectancy under reasonable conditions of from 1½ to 5 years. Their usefulness under extreme conditions has been as short as 6 months and as long as 12 years.

For most ion exchange resins it becomes impractical, from a cost and efficiency basis, to continue operating with resins which have lost in excess of approximately 30% of their exchange capacity. Generally, the replacement point for cation resins is when their cross-linkage has been attacked and their moisture content has gone up drastically. With common 8% cross-linked resin (in the sodium form), a moisture content of above 50% signals that replacement is needed.

Replacement is, of course, dependent on the resin and regenerant costs, but also on downtime for regeneration due to short service cycles. If capacity loss can be traced to foulants, then treatment to clean up the resins and recover capacity can be an economical step to take, and can possibly be instituted as a regular maintenance procedure.

In general, when ion exchange resins themselves cause problems in performance of an ion exchange system, these problems seldom occur rapidly or over a short period of time. Reduction in capacity usually occurs over a period of months or years unless provoked by unusual operating procedures, fouling, or mechanical malfunction.

Generally, resin deterioration or losses are divided into two categories, physical and chemical.

Physical

- Breakdown of resin beads or granules into small-size fragments by mechanical attrition (in backwash or hydraulic transfer), thermal shock, osmotic shock, or water hammer.
- Loss of resin due to backwash at high flowrates, broken distributors, gas or air flotation, or clumping.
- Loss of resin in service cycle due to broken collector or upset underbedding.
- High pressure drop across the resin bed and the resin interface with the distributor surface, which would cause crushing of the resin beads with subsequent loss of fines in the backwash.
- Faulty operation.
- Improper maintenance of equipment or cleaning of resins.

Chemical

- Oxidation attack by substances such as chlorine, chlorates, oxygen, or ozone, accelerated by catalytic effect of trace quantities of iron or copper.
- Loss of capacity because the thermal stability of the reactive groups has been exceeded. On the anion exchange resins in particular, this will reduce their basicity or destroy the reactive group.
- Fouling the resins with precipitates or organics so that they require cleanup procedures, and in some cases to the extent where cleanup may not be economically feasible.
- Improper control of regenerant chemicals, both in amounts and concentration.
- Use of inferior-quality chemicals.

Equipment

Like resins, ion exchange system equipment also eventually reaches the limits of its durability. This is particularly true of the components with moving parts or instrumentation, all of which need maintenance on a continuing or periodic basis.

Monitoring instruments such as conductivity, pH, specific ion recorders, temperature indicators, flow meter, level controls, and others should be checked and calibrated frequently. Regeneration chemical pumps should be checked for delivery rates as well as for the proper concentration of diluted chemicals periodically to the resin tanks. Leaking valves can be a cause for contamination of treated water. Sampling ports on critical lines may be useful in spotting such trouble.

Some parts of equipment, such as the distributor systems, are difficult to inspect without shutting down the system and opening the tanks. These parts are a source of many problems, but they do produce their own symptomatic signs. Here are some warnings of trouble:

Underbedding, bottom distributor, or collector—resin leakage in the service cycle, uneven surface of resin bed during or after backwash, early breakthrough of end point or short service cycle, and poor regeneration.

Plugged bottom distributor—high pressure drop in service cycle and poor regeneration due to channeling.

Broken top distributor—resin loss in backwash, short service cycle, and disturbed top of resin bed.

Plugged top distributor—short service cycle, channeling (since not all of distributor will be plugged and the remaining flow is directed at high velocity to a small section of the resin bed area), high pressure drop, and disturbed top of the resin bed.

Broken regenerant distributor—short service cycle, channeling of regenerant only to portions of resin bed, faster regenerant draw due to less back pressure, and low concentration of regenerant.

Plugged regenerant distributor—regenerant draw slow due to back pressure (so that it may not be completed in timed sequence), high concentration of regenerant, and short service cycle.

Water hammer—This is a subject that deserves more in-depth discussion in view of its frequent occurrence. Sudden surges or change in water flow produced by fast-acting valves, booster pumps being activated, or sudden interruption of flow can all be sources of water hammer. This effect may be familiar to many people who have heard the knocking and rattling noise of water pipes in homes when faucets were rapidly opened or closed. The pressure waves created by this phenomenon are enormous. The pressures developed are dependent on the linear flow velocity and pipe length, and vary inversely with the valve closure time. Instantaneous pressures as high as 800 to 900 psig can be created by this water hammer effect. This sudden surge can be sufficient to cause serious bending and breakage in piping and internals in resin tanks, as well as crushing of ion exchange resins into powder.

Results of water hammer have been observed in several ion exchange systems by the author. In one case, the results were: resin loss, with the fines being removed in backwash, and resulting low resin volume; cation exchange resin fines being found in the following anion exchange resin tank; poor conductivity of the product water resulting from sodium leakage from the anion unit due to the sodium form of the cation resin in the anion tank with caustic regeneration; and silica leakage higher than normal. This is an example of how one problem can produce a variety of symptoms. This could have lead one to the conclusion that the resin had poor physical strength, while in reality the resins were normal and the problem originated with the mechanical problem of water hammer. The physical forces of this water hammer were strong enough to put significant bends and distortions in a 3-inch diameter stainless steel distributor.

Resin tank linings—Another equipment problem (limited now, but likely to become more frequent as the ages of the installed systems become greater) is the lining used in deionization tanks and systems. Since the lining materials used are organic (rubber, saran, PVC, etc.) the long-term durability is dependent on the exposure conditions to oxygen, chlorine, ozone, or other oxidants, as well as to oils or organic solvents in combination with the temperature of operation.

Some leaching, erosion, and abrasion can take place on the lining surface to liberate the following lining components: plasticizers, fine-size fillers (carbon black, titanium dioxide, clays, color pigments, etc.), antioxidants, stabilizers, and organic products of the degradation of the polymers. These contaminants may not become problems until the system is 15 to 20 years old. At that time, however, such breakdown may be significant for systems where ultrahigh-quality product water is required. Particulate and organic foulants that are not present in the water being treated may be found in the produced water from this source.

Regenerant eductors and pumps—A frequent cause of low run length or capacity is the supply of regenerant by eductors or pumps that, due to corrosion, wear, plugging, or lack of proper maintenance, are not providing the designed flow of chemicals. This will reduce the amount of regenerant and/or reduce the concentration of regenerant to the resin. Back pressure on the regenerant supply caused by a plugged distributor or a compacted resin bed may also be a factor in shorting the regenerant flow through the resin bed during a timed regenerant chemical introduction period.

Operations

It has been repeated numerous times in these discussions how many variables there are to cause problems in ion exchange systems. It is time now to consider the variables who control the systems: the operators and supervisors responsible for the operation of the systems. Most of the direction of this book has been to these people, who have the responsibilities to see that the ion exchange systems perform to their best capabilities. Human error is always a factor to be dealt with in any operation. If the operator understands the principles of the system, he will be better prepared to follow the instructions and to make the proper choices when they are called for.

Many problems arising in ion exchange system operations are due to individuals making changes contrary to information given in the operating instructions, without consulting with the equipment supplier or engineering personnel. Frequently not only one, but several changes are made at a time, which can further complicate the operation of the system.

If changes are to be made, try to make only one change at a time and then review the results of that change over at least three or more cycles of operation. It takes at least that long for the ion exchange system to equilibrate, particularly when changes are made in regenerant dosage levels or conditions. All changes should be noted in writing so that other operating personnel are aware of them and will follow through with their evaluation.

Record Keeping
Record keeping is the most valuable tool to define and locate sources of problems that may occur. The records kept will of course be dependent on the type of system or systems installed. It is suggested that at least the following records should be kept for each complete operation cycle, with additional observations recorded of any unusual occurrences (such as shutdowns, malfunction of equipment, instrumentation, leaks, etc.) that take place.

For softeners
Raw water - Hardness
Regeneration level
Rinse gallonage
Final effluent - Hardness, run length in gallons

For two-bed DI SAC + WBA
Raw water - Hardness, TMA
SAC - Regeneration level, rinse gallons
WBA - Regeneration level, rinse gallons
Final effluent - pH average, conductivity average, run length in gallons

For SAC + SBA
Raw water - Hardness, TMA, M.O. alkalinity, silica
SAC - Regeneration level, rinse gallons
SBA - Regeneration level, rinse gallons
Final Effluent - pH, conductivity, silica, run length in gallons

For mixed-bed SAC/SBA
Same as for SAC + SBA above

The above gives some ideas as to the records that might be kept, adjusted as required for other systems. For systems with degassing equipment, it would be desirable as a check on their efficiencies to monitor the carbon dioxide level out of the forced-draft degasifier. and possibly the carbon dioxide and oxygen out of a vacuum degasifier. It may be helpful to run pH out of SAC service run to determine pH rise indicating overrun on SAC resin. Measuring pH as well as conductivity out of WBA may be helpful to check the end point of run by pH drop.

It is also helpful to maintain a book which contains: a flow diagram of the system; a copy of the original operating manual; information on the types and volumes of the resins used and when they were installed or replaced; references to end points of rinses and service cycles; and a list of mechanical modifications made to the system and the dates they were made.

From the records kept, it will be easy to check whether changes have occurred gradually over a long period of time (which might indicate resin problems) or suddenly (which would be an indication of mechanical failures, water composition variation, faulty operation, or regeneration changes). Organic fouling may best be indicated by a rise in conductivity and a drop in pH and probably a lengthening of the rinse of the SBA resin.

The early treatment of any problem is important to prevent major shutdown of the system as well as to make cleanup, in case of fouling, more easily achieved. The longer foulants accumulate, the more difficult it is to remove them.

It should also be remembered that if there are significant variations in the service cycle run length, and if a long run can be obtained after several short runs with the raw water quality constant, then the problem must be mechanical in nature. With consistent regeneration levels and conditions, the resin would not have recovered capacity once it had been lost.

An interesting problem that was solved by keeping good records was one where the record book showed an increasing frequency of SBA resin replacement over a period of years, until the life appeared to be unusually short. It was finally resolved by an alert engineer who reviewed the information in the record book. He found that the problem was in the cation resin, which had become progressively decrosslinked by oxidation attack, to the point where there were polymer fragments of sulfonated styrene sloughing off the SAC resin and irreversibly fouling the SBA resin. When the SAC resin was replaced, the SBA resin replacements were reduced to normal.

The detective work of troubleshooting (or preventing serious problems) is materially aided by a well-kept records book, where the information has been faithfully and honestly entered and followed up by frequent review. Where there has been long-term experience that, for instance, the raw water has not changed appreciably, then the frequency of entering data may be extended to longer time intervals. In such a case it need not be done for each complete operation cycle or each day, but it should not be completely eliminated.

In one ion exchange system where the feedwater had been running very consistent over a period of years, there was rather suddenly a problem with short runs of the SAC resin in the DI system. It was traced to a change in the flocculant use in the pretreatment by the municipal water plant (which supplied the feedwater). Excess flocculant had been used, which resulted in flocculant fouling of the cation resin. Cleanup of the flocculant-fouled SAC resin was finally achieved (after laboratory tests) by hot caustic treatment of the regenerated SAC resin, followed by double acid regeneration before the resin was put back in service.

Troubleshooting Aids
There are a number of excellent and extensive charts, tables, and detailed manuals available on troubleshooting ion exchange resin systems; but because

of the numerous variables and interrelated factors, it is not easy to come up with a presentation that can cover all facets of all problems. Herein we try a somewhat different approach, based on the way medical doctors diagnose illnesses in people, or perhaps even better related to the way veterinarians diagnose illnesses in animals who can not communicate their problems to the doctor.

We do have an advantage, however, in having the patient (the ion exchange system) already wired up or connected to instrumentation devices that are, or should be, giving us signs of the health of the patient, or the symptoms to assist in our diagnosis. In most cases, these available indicators are not fully utilized to recognize potential problems that may show up later to effect final water quality or quantity. Unfortunately, the usual sequence of events is to concentrate more attention on the final water quality. Only when signs of disaster appear (the patient is dying), do we start looking frantically for sources of the problem.

A patient in a hospital has all vital signs recorded regularly on his charts. So must the operator of ion exchange systems keep records and then have the knowledge to interpret the changes that occur so that action can be taken to correct a problem at its source, before complications develop. As in the medical profession, continued training in interpreting the symptoms, knowing where to look for the sources of the problem, and knowing how to take corrective action are vital necessities.

Maintenance of the instrumentation, the equipment, and the ion exchange resins are also fundamental to the proper operation of ion exchange systems and are often overlooked until complete failure occurs. Maintenance is best handled by frequent and regular inspections and testing.

The tables that follow are presented to outline only a few of the very many sources and symptoms of problems that can manifest themselves in the final product water quality, quantity, and cost. There are certainly many additions that can be made to the list presented herein. Anyone who considers the number of possible combinations of problem sources may be staggered by these tables when attempting to use them as a way to solve a particular problem. At any rate, they certainly suggest a lot of places to look for trouble.

The cross-referencing of these tables with some of the information presented previously in this text is intended to make them useful tools in troubleshooting and as a warning of possible sources of problems. (Refer to the Subject Index for page numbers.) The problem sources listed include items which, since they require continual observation and maintenance by the operators, might be included in a check list of things to do.

We have not gone into the many detailed procedures that can be used in correcting the problems once the sources are located, as that would in itself be the basis for a book.

TABLE 9-A
Problems or Symptoms

Quality	Quantity	Cost
Conductivity	Capacity	Regenerants
pH	Pressure drop	Resins
Particulates		Water
Microorganisms		Waste
Specific ions		Equipment
Organics		Electric power
		Manpower

TABLE 9-B
Problem Sources—Components

Water	Ion Exchange Resins	Equipment
Pretreatment	Cation quality	Tanks
TDS	Anion quality	Tank linings
TMA	Regenerants	Distributors
TOC	Shock	Instrumentation
TSS	thermal	Valves
pH	osmotic	Regenerant supply
Colloids	hydraulic	Pumps
Ion ratios	Clumping	Level controls
Flowrate	Temperature	
Gases	Resin loss	
Organics	Separation	
Oxidants	Resin mix	
Oil	Cleaning	
Temperature	Precipitation	
Polyelectrolytes	Foulants	
Multivalent ions	Fines	
Alkalinity	Resin ratios	
Microorganisms		

Operations
Record keeping
Operating instructions
Observations
Maintenance
Manpower changes
Human error
Training

TABLE 9-C
Problem Sources in Operation Cycles

Backwash & Separation	*Regeneration*	*Displacement Rinse*
Temperature	Regenerant	Flowrate
Gases	purity	Temperature
Oil	concentration	Water quality
Clumping	mixing	Time
Classification	level	
Quality BW water	temperature	
MB separation	flowrate	
Particulate removal	Precipitation	
	Multivalent ions	
	Organics	
	Polyelectrolytes	
	Resin level	
	MB resin ratio	
	Shock	
	thermal	
	osmotic	
	Dilution water quality	
	Eductor or pumps	
	Channeling	
Fast Rinse	*Service*	*MB Air Mix*
Flowrate	Flowrate & changes	Air volume
End-point control	Water temperature	Time
Organics	Gases	Excess water
Regenerant hideout	Degassing operation	
	Channeling	
	Pressure drop	
	Valve leaks	
	Color throw	
	Conductivity	
	pH	
	Run length	
	Particulates	
	Microorganisms	
	Organics	

When the three criteria of quality, quantity, and cost of produced water are met for any ion exchange system, there would not appear to be any problem in the system. However, meeting these criteria on the product water does not eliminate the possibility that there may be some imbalance or malfunctions in the operation of units or systems upstream. This may occur because the final units are able to handle the extra or changed load for some period of time and still produce good product water. This is the reason why good preventive-maintenance troubleshooting is necessary to catch these changes before they finally show up in the product water.

The three criteria for product water and the measurements or values used to detect problems or symptoms of problems that may require troubleshooting are shown in Table 9-A. The sources of problems and the places where one would look for them are shown in Table 9-B for the components of ion exchange systems.

The remaining sources of problems are those associated with the operation cycles of the ion exchange resin systems, and are given in Table 9-C.

In the above tables, there are a total of 114 potential sources of problems under 12 major headings listed. Many of them can be interrelated and together cause a specific problem. Interestingly, the most frequently indicated source of problems listed are:

- Mechanical or physical factors
- Organics of all varieties
- The ionic composition of water and changes
- Ion exchange resins
- Regenerants
- Particulates of all varieties
- Temperature factors

Troubleshooting experiences in the field tend to support this general order of the sources most frequently found to be the cause or causes of problems in ion exchange systems.

CHAPTER 10

WASTE

In water treatment, ion exchange resins are used as a means of replacing unwanted ions with desirable ions and concentrating the unwanted ions on the resins to later be removed in a concentrated waste by chemical regenerants. This waste must be disposed of (in most cases) in keeping with the restrictions placed by the sewage districts in which the ion exchange facility is located. This problem has become of greater importance as the environmental controls have been tightened and the sewage treatment facilities have become more sophisticated in their handling of wastes.

The first and most obvious problem is that the disposal lines or mains must be protected from corrosion and from precipitation of sludges. There are therefore requirements of pH control to prevent corrosion on the low pH or acidic side and precipitation on the high pH or caustic side.

There are also limitations on the TDS or salt content of the waste, with cost penalties (in some cases) being assessed for high TDS. These assessments are justified in view of the potential recovery of water from the sewage treatment process for disposal into freshwater streams, flooding basins, agricultural use, or injection into groundwater reservoirs.

The more efficient use of regenerating chemicals is not only important to the cost of operation, but also will reduce the amount of excess chemicals that must be neutralized (with added chemicals) and also the TDS of the waste. If the average pH of the waste is low (that is, if there is more excess acid from the regeneration of the cation resin than excess caustic from the regeneration of the anion resin), then caustic (or less expensive lime or soda ash) must be added to bring the pH up to the acceptable range before the waste can be dumped to the sewer. If this is the case, then it might be well to consider increasing the caustic regeneration level of the anion resin to take advantage of better regeneration. This also results in better capacity and quality, with the excess caustic producing a more neutral waste. Of course, the reverse action might be taken when the average pH is too high in the combined waste from cation and anion regeneration.

One commonly overlooked problem in the combined waste of cation and anion regeneration is that when the pH of the cation waste is raised much above pH 7 (to pH 8 or over) by the basic anion waste, several precipitates can form. They include calcium and magnesium hydroxide and, most troublesome, the apparent formation of the very insoluble magnesium silicate ($MgSiO_3$), which will not readily redissolve even when the pH is brought back below pH 7 by acid addition. By adding the anion basic waste to the cation acid waste and not allowing the pH to go above 8, this precipitate may be prevented from forming. If the anion basic waste exceeds the amount of cation acid waste, then additional acid must be added so that the combined waste will stay below pH 8.

It is possible to design ion exchange systems to a water analysis (hopefully a fairly consistent one) where the waste pH will come out very close to neutral, and the capacities of the anion and cation resins will be balanced. The use of WAC and WBA resins in the system may aid in achieving this goal if excess regenerant chemicals from the SAC and SBA resins are used to regenerate these weak electrolyte resins and obtain additional capacity (if the water analysis is favorable).

Conserving water not only makes sense from a cost standpoint, but also in terms of reducing the waste load. There are several ways that this can be achieved: recycle rinses, reclamation of backwash water, and recycling of some portion of the displacement rinse as dilution water for the chemical regeneration.

It is also possible to recycle a portion of the chemical regenerants, which could be used as the first portion of the regeneration in the next cycle. Countercurrent regeneration can materially reduce both water and chemical waste because of the better efficiencies in the countercurrent operation.

Many of these conservation techniques require additional costs for resins and equipment (pumps, storage tanks, valves, instrumentation, and piping), but the payoff in operating costs (reduced chemicals, water, waste disposal) can in most cases result in a return on the investment. Calculations of the probable maximum potential savings that could be achieved in the ideal countercurrent ion exchange water treatment plant as compared to an average co-current regeneration plant would be a 30% reduction in chemicals and a 75% savings in water and waste volume. These savings would of course be dependent on the analysis of the feedwater being treated and the quality of produced water required.

The biggest shortcoming of the ion exchange water treatment process is that the waste created is the combination of the ions removed from the water and the chemicals used to regenerate those ions from the ion exchange resins. Membrane processes such as reverse osmosis use electrical energy to remove the ions from the water (with pressure in the case of RO), and as a result the waste brine being produced is almost entirely composed of the ions in the water being treated (except for additives used to control scaling or fouling). Fortunately for the ion exchange resin process, its ability to produce the highest quality of water in large volume from a fairly broad range of ionic impurities is quite efficient in its chemical usage.

There are a few methods of recovering regenerant chemicals from waste brines, but most are not very efficient, and they require setting up another system using more energy and other chemicals.

IN CLOSING

Mention was made earlier in this book of some of the other very interesting and surprisingly large volume applications for ion exchange technology. Although most of these are involved with systems where water is the main carrier or solvent, there are a number of applications where there is no water present and nonaqueous liquids are used. Many of the same basic water treatment principles that we have discussed are involved with these and other applications. There are, however, whole new groups of terminolgy used.

It has been said that ion exchange principles are involved in practically everything we can see, touch, or eat. This is what makes it a most fascinating subject.

Postscript
The information and discussions presented in this book will not produce instant experts in ion exchange, but it is the author's hope that it will lead to the reader's better understanding of ion exchange. It should allow him to go on to learn more in this field from the many excellent texts available, and to become as well a more productive user of this technology and its applications.

GLOSSARY

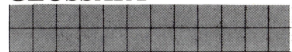

ACID—the hydrogen form of an anion.

AFFINITY—the force that causes two things to combine.

ALKALINITY—(abbr. Alk) - bicarbonate, carbonate, or hydrate amounts in water. Can be expressed as "M" Alk to a methyl orange titration end point or "P" Alk to a phenolphthalein end point.

ANALYSIS—separation and measurement of component parts.

ANION—an ion of negative charge attracted to the anode.

AQUEOUS—using water as a solvent.

ATTRITION—breakage or wear of resins.

BACK PRESSURE—pressure exerted against a flow.

BACKWASH—(abbr. BW) - upward flow of water through a resin to cleanse, expand, and classify the resin.

BASE—the hydroxyl form of a cation or a compound that can neutralize an acid.

BED DEPTH—measure of resin bed from top to bottom.

BED VOLUME—(abbr. BV) - the amount of space occupied by the resin bed in cubic feet or gallons.

BREAKTHROUGH—see END POINT.

BRINE—a salt solution, generally sodium chloride.

CAPACITY—TOTAL - the ultimate exchange capacity of the resin.

OPERATING—the portion of total capacity utilized in practical ion exchange operation.

SALT-SPLITTING—the portion of total capacity to split neutral salts.

CATION—an ion of positive charge attracted to the cathode.

CHANNELING—the flow of water or solution taking the "line of least resistance" through a resin bed.

CLASSIFICATION—by backwash to obtain a resin bed which is graduated in resin size from coarse on the bottom to fine on the top.

COAGULATION or COALESCE—to bring together small particles into a single larger mass which can be filtered or settled out.

COLLOIDAL—composed of extremely small size particles which are not removed by normal filtration.

COLOR THROW—the release of color from an ion exchange resin on soaking or on being used in water treatment.

CONDENSATE POLISHERS—ion exchange resins being used to remove or exchange ions as well as to filter condensate for reuse in the steam cycle.

CONDUCTIVITY—the ability of electric current to flow through water as a measure of its ion content in mhos or micromhos.

CROSS-CONTAMINATION—intermixing of one resin with another of opposite charge or of two water streams.

CROSS-LINKAGE—the connection between two or more polymer chains to tie them together, as is done by DVB.

DAY TANKS—tanks used to hold diluted regenerant chemicals prior to being educted or pumped to the resin bed.

DEGASIFIER or DEAERATOR—a system to remove dissolved gas or gases from water.

DEIONIZE—(abbr. DI) - to remove ions from solution by means of ion exchange resins.

DENSITY—weight per unit of volume.

DISSOCIATE—the process of ionization of an electrolyte or a salt upon being dissolved in water, forming ions of cation and anion.

DIVINYLBENZENE—(abbr. DVB) - a difunctional monomer used to cross-link polymers.

DYNAMIC—an ion exchange reaction taking place as the water moves past ion exchange resins.

EDUCTOR—a device that by flow of water through it creating a vacuum, draws a solution into the water stream.

Glossary

EFFLUENT—flow of water out of a tank or system.

ELECTRODES—conductive materials, placed in water solution, which have a positive or negative charge.

ELECTROLYTES—materials that when dissolved in water form ions which are capable of carrying electric current.

ELECTRONEUTRALITY—where the positive charges equal the negative charges.

ELUTION—to remove ions from a resin by passing other ions in solution of higher concentration or affinity through the resin bed.

END POINT—a preselected value or measurement used to determine when the service cycle is completed or that a breakthrough has occurred.

EQUATION—an expression of a reaction using chemical symbols.

EQUILIBRIUM—the stage in a reversible reaction when there is no driving force in either direction.

EQUIVALENT—equal valence.

EXCHANGE SITES—reactive groups on an ion exchange resin.

EXOTHERM—giving off heat in a reaction or solution of a substance.

EXTRAPOLATION—extending information beyond the point where proven data is available.

FINES—resin particles smaller than 50 mesh.

FLOCCULANTS—materials which can form gelatinous clouds of precipitate to enclose fine particles of suspended dirt to settle them from the water.

FMA—(abbr.) free mineral acidity, or sum of the mineral acids. See also TMA.

FORCED-DRAFT DEGASIFIER—a tower in which water droplets descend through a flow of air blown upwards to remove gases such as carbon dioxide.

FOULANTS—substances which coat or adsorb onto and absorb into ion exchange resin to reduce available capacity.

FREE BASE—the regenerated form of a weak base anion resin.

FREEBOARD—the space above the resin bed to accommodate resin expansion in backwash.

FREE CO_2—dissolved carbon dioxide gas in water.

FUNCTIONAL GROUPS—see EXCHANGE SITES.

GELULAR—a jellylike appearance with no apparent physical pores.

GRAVIMETRIC—measurement by weight.

HARDNESS—the combination of calcium and magnesium in water. Can include iron and other multivalent cations.

HEAD LOSS—the drop in pressure of a water flow through a resin bed.

HYDRAULICS—a movement or action resulting from liquid flow.

HYDROLYSIS—the splitting of a salt and water into its ions and formation of a weak acid or base, or both. As in ion exchange:
$$R^-NH_4^+OH^- \rightarrow R^-H^+ + NH_4^+OH^-$$
where R^- is a WAC resin.

HYDROMETER—a device to measure specific gravity of fluids.

HYDROXYL—the anion of water or OH^-, also present in all hydroxides.

INERT MEDIA—a resin with no reactive groups.

INFLUENT—a liquid flow into a resin tank or system.

INTERFACE—the surface between two resins, or resin and a distributor.

INTERLOCK—a device which will prevent one action from happening while another action is proceeding.

ION EXCHANGE—the interchange of one ion in solution with another ion on an insoluble material.

JACKSON UNITS—a measure of turbidity or suspended materials by optical obstruction of light rays passed through a water sample, as compared to a standard.

KINETICS—the effects of forces on the motion of matter, of concern with the velocity or speed of reactions.

LAYERED or STRATIFIED BED—resins with sufficient difference in density and hydraulic characteristics to be layered in the same tank in place of two separate tanks.

Glossary

LEAKAGE or SLIPPAGE—the amount of an ion or ions coming from a resin bed during the service cycle.

MACROPOROUS—having large pores.

MOISTURE CONTENT—the water loss of a fully hydrated resin under controlled drying conditions.

MOLECULAR WEIGHT—the relative weight of a molecule on a scale where oxygen has a given weight of 16.

MOLECULE—the smallest particle of an element or compound retaining its characteristics.

MONOMER—a single reactive molecule capable of combining with another like itself or another different monomer to form a polymer. Where two different monomers combine, the resulting polymer is called a copolymer.

NEUTRAL—in pH terms, 7; neither acid nor basic.

NONREACTIVE SILICA—polymerized or colloidal silica.

OHM—a unit of resistance to the passage of electric current.

ORGANICS—carbon-containing compounds, generally from vegetation or nonmineral origin.

OSMOSIS—the process where water on one side of a membrane tries to pass through the membrane to dilute a salt solution on the other side in an attempt to equalize the concentration of salt on both sides of the membrane.

pH—a number system that relates to the acidity, neutrality, or basicity of water.

POLYELECTROLYTE—a high-molecular-weight water-soluble polymer with high charge density or a multicharged ion.

POLYMER—see MONOMER.

POLYMERIZATION—the act of reacting monomers to form a long chain (POLYMER).

POROSITY—the degree of openness of a spongelike quality in a resin structure.

POTABLE WATER—meets drinking-water quality standards.

PRECIPITATION—forming an insoluble substance from a combination of soluble ions.

PRESSURE DROP—a loss in pressure due to flow restrictions or friction.

PRETREATMENT—includes flocculation, settling, filtration, or any treatment that water has received prior to ion exchange.

PROCESS WATER—water used in a product or in manufacture of a product.

REACTION EQUATION—the expression of the exchange of ions brought about by differences, affinities, concentrations, temperature, flow, and other influences.

REGENERANT—a chemical used to restore the ion exchange resin back to its desired form.

REGENERATION LEVEL—the amount of regeneration used, usually in pounds of 100% form of the chemical.

RESISTANCE—the property of opposing the flow of electric current.

REVERSE OSMOSIS—to reverse the flow of water through a membrane by applying pressure to overcome the osmotic pressure.

SAC—(abbr.) strongly acidic cation resin.

SALT-SPITTING—to replace the cation (or anion) of a neutral salt such as sodium chloride (NaCl) with hydrogen (or hydroxyl) ion.

SATURATED—the maximum amount of a substance that can be put into solution (maximum capacity).

SBA—(abbr.) strongly basic anion resin.

SELECTIVITY—the difference in attraction of one ion over another by an ion exchange resin.

SEQUENTIAL—one action occurring followed by others in a given order, as opposed to simultaneous actions.

SHORT CIRCUIT—see CHANNELING.

SOFTENING—the process of replacing hardness with sodium by cation exchange.

SPACE VELOCITY—the flow per unit volume of resin or gpm/CF.

SPECIFIC GRAVITY—the ratio of the weight of a given volume of a substance to the weight of an equal volume of water at a fixed temperature.

Glossary

SPECIFIC RESISTANCE—the resistance between opposite faces of a one-centimeter cube of a given substance and expressed as ohms-cm.

STATIC—an ion exchange reaction occurring with a volume of liquid in continuous contact with a volume of resin.

STRAINER—a slotted or screened sieve to filter a flowing stream of water.

STRONG ELECTROLYTE RESIN—the equivalent of strongly acidic or strongly basic resins and capable of splitting neutral salts.

SUPERFICIAL LINEAR VELOCITY—flow of water per unit of area or gpm/ft^2.

SUPERSATURATED—to contain more in solution than normal for a given temperature.

SUPPORT MEDIA—a graded-particle-size, high-density material such as gravel, anthrafil, quartz, etc. used to support the resin bed.

TDS—(abbr.) total dissolved solids. TDS = TS-TSS.

TITRATE—to determine the amount of a substance in solution by adding a measured amount of another substance to produce a desired reaction or end point.

TMA—(abbr.) total mineral acidity or anions.

TOC—(abbr.) total organic carbon.

TRAIN—a single ion exchange system capable of producing the treated water desired, such as SAC + SBA, with multiple trains being duplicates of the single system.

TS—(abbr.) total solids. TS = TDS+TSS.

TSS—(abbr.) total suspended solids. TSS = TS-TDS.

TURBIDITY—see JACKSON UNITS.

UNDERBEDDING—see SUPPORT MEDIA.

VACUUM DEGASIFIER—a tower to which a vacuum is applied and through which water droplets descend to remove dissolved gases.

VALENCE—is measured by the number of atoms or ions of hydrogen it takes to combine with or be replaced by an element or radical. In short, the number of positive or negative charges of an ion.

VOID VOLUME—the space between particles of ion exchange resins in a settled bed; also called interstitial volume.

WAC—(abbr.) weakly acidic cation resin.

WASTE WATER—the unrecycled portion of the water used in the backwash, regeneration, and rinse cycles.

WATER HAMMER—instantaneous surges of water pressure caused by sudden interruptions in water flow in pipe or tank systems.

WATER SOFTENING—to exchange sodium for the hardness in water by ion exchange.

WBA—(abbr.) weakly basic anion resin.

WEAK ELECTROLYTE— the equivalent of weakly acidic or weakly basic resins not capable of splitting neutral salts.

ZEOLITE—a mineral composed of hydrated silicates of aluminum and sodium or calcium. The term has been used, sometimes improperly, to describe softening done by synthetic ion exchange resins.

TABLES AND CONVERSION FACTORS

The tables and figures given on the following pages have been taken from *"Engineering Manual for the Amberlite® Ion Exchange Resins."* They are reproduced by permission from Rohm and Haas Co., Philadelphia, PA.

ELECTRICAL RESISTANCE OF DILUTE ELECTROLYTES (Deionized Water at 25°C.)

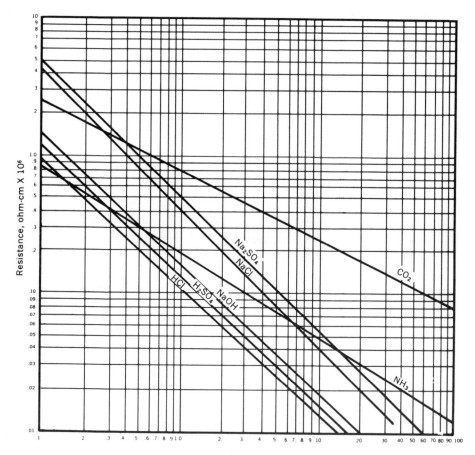

Conductivity Micromhos-cm. @25°C	Resistivity Ohms-cm. @25°C	Dissolved Solids Parts per Million (ppm)	Approximate Grains/Gallon (GPG) as $CaCO_3$
0.056	18,000,000	0.0277	0.00164
0.059	17,000,000	0.0294	0.00170
0.063	16,000,000	0.0313	0.00181
0.067	15,000,000	0.0333	0.00193
0.072	14,000,000	0.0357	0.00211
0.077	13,000,000	0.0384	0.00222
0.084	12,000,000	0.0417	0.00240
0.091	11,000,000	0.0455	0.00263
0.100	10,000,000	0.0500	0.00292
0.111	9,000,000	0.0556	0.00322
0.125	8,000,000	0.0625	0.00368
0.143	7,000,000	0.0714	0.00415
0.167	6,000,000	0.0833	0.00485
0.200	5,000,000	0.100	0.00585
0.250	4,000,000	0.125	0.00731
0.333	3,000,000	0.167	0.00971
0.500	2,000,000	0.250	0.0146
1.00	1,000,000	0.500	0.0292
1.11	900,000	0.556	0.0322
1.25	800,000	0.625	0.0368
1.43	700,000	0.714	0.0415
1.67	600,000	0.833	0.0485
2.00	500,000	1.00	0.0585
2.50	400,000	1.25	0.0731
3.33	300,000	1.67	0.0971
5.00	200,000	2.50	0.146
10.0	100,000	5.00	0.292
11.1	90,000	5.56	0.322
12.5	80,000	6.25	0.368
14.3	70,000	7.14	0.415
16.7	60,000	8.33	0.485
20.0	50,000	10.0	0.585
25.0	40,000	12.5	0.731
33.3	30,000	16.7	0.971
50.0	20,000	25.0	1.46
100.0	10,000	50.0	2.92
111	9,000	55.6	3.22
125	8,000	62.5	3.68
143	7,000	71.4	4.15
167	6,000	83.3	4.85
200	5,000	100	5.85
250	4,000	125	7.31
333	3,000	167	9.71
500	2,000	250	14.6
1,000	1,000	500	29.2
1,110	900	556	32.2
1,250	800	625	36.8
1,430	700	714	41.5
1,670	600	833	48.5
2,000	500	1,000	58.5
2,500	400	1,250	73.1
3,330	300	1,670	97.1
5,000	200	2,500	146
10,000	100	5,000	292

VOLUME OF VERTICAL CYLINDRICAL TANKS

Diameter	U. S. Gallons per 1' of Depth	Area in Sq. Ft. Cu. Ft. per 1' of Depth	Diameter	U. S. Gallons per 1' of Depth	Area in Sq. Ft. Cu. Ft. per 1' of Depth	Diameter	U. S. Gallons per 1' of Depth	Area in Sq. Ft. Cu. Ft. per 1' of Depth
1'	5.87	0.785	6'	211.5	28.27	28'	4606.	615.8
1' 1"	6.89	0.922	6' 3"	229.5	30.68	28' 6"	4772.	637.9
1' 2"	8.00	1.069	6' 6"	248.2	33.18	29'	4941.	660.5
1' 3"	9.18	1.227	6' 9"	267.7	35.78	29' 6"	5113.	683.5
1' 4"	10.44	1.396	7'	287.9	38.48	30'	5288.	706.9
1' 5"	11.79	1.576	7' 3"	306.8	41.28	31'	5646.	754.8
1' 6"	13.22	1.767	7' 6"	330.5	44.18	32'	6016.	804.3
1' 7"	14.73	1.969	7' 9"	352.9	47.17	33'	6398.	855.3
1' 8"	16.32	2.182	8'	376.0	50.27	34'	6792.	907.9
1' 9"	17.99	2.405	8' 3"	399.9	53.46	35'	7197.	962.1
1' 10"	19.75	2.640	8' 6"	424.5	56.75	36'	7616.	1018.
1' 11"	21.58	2.885	8' 9"	449.8	60.13	37'	8043.	1075.
2'	23.50	3.142	9'	475.9	63.62	38'	8483.	1134.
2' 1"	25.50	3.409	9' 3"	502.7	67.20	39'	8940.	1195.
2' 2"	27.58	3.687	9' 6"	530.2	70.88	40'	9404.	1257.
2' 3"	29.74	3.976	9' 9"	558.5	74.66	41'	9876.	1320.
2' 4"	31.99	4.272	10'	587.5	78.54	42'	10360.	1385.
2' 5"	34.31	4.587	10'6 "	647.7	86.59	43'	10860.	1452.
2' 6"	36.72	4.909	11'	710.9	95.03	44'	11370.	1521.
2' 7"	39.21	5.241	11'6"	777.0	103.9	45'	11900.	1590.
2' 8"	41.78	5.585	12'	846.0	113.1	46'	12430.	1662.
2' 9"	44.43	5.940	12' 6"	918.0	122.7	47'	12980.	1735.
2' 10"	47.16	6.305	13'	992.9	132.7	48'	13540.	1810.
2' 11"	49.98	6.681	13' 6"	1071.	143.1	49'	14110.	1886.
3'	52.88	7.069	14'	1152.	153.9	50'	14690.	1964.
3' 1"	55.86	7.467	14' 6"	1235.	165.1	52'	15890.	2124.
3' 2"	58.92	7.876	15'	1322.	176.7	54'	17130.	2290.
3' 3"	62.06	8.296	15' 6"	1412.	188.7	56'	18420.	2463.
3' 4"	65.28	8.727	16'	1504.	201.1	58'	19760.	2642.
3' 5"	68.58	9.168	16' 6"	1600.	213.8	60'	21150.	2827.
3' 6"	71.97	9.621	17'	1698.	227.0	62'	22580.	3019.
3' 7"	75.44	10.08	17' 6"	1799.	240.5	64'	24060.	3217.
3' 8"	78.99	10.56	18'	1904.	254.5	66'	25500.	3421.
3' 9"	82.62	11.04	18' 6"	2011.	268.8	68'	27170.	3632.
3' 10"	86.33	11.54	19'	2121.	283.5	70'	28790.	3848.
3' 11"	90.13	12.05	19' 6"	2234.	298.6	72'	30450.	4072.
4'	94.00	12.57	20'	2350.	314.2	74'	32170.	4301.
4' 1"	97.96	13.10	20' 6"	2469.	330.1	76'	33930.	4536.
4' 2"	102.0	13.64	21'	2591.	346.4	78'	35740.	4778.
4' 3"	106.1	14.19	21' 6"	2716.	363.1	80'	37600.	5027.
4' 4"	110.3	14.75	22'	2844.	380.1	82'	39500.	5281.
4' 5"	114.6	15.32	22' 6"	2974.	397.6	84'	41450.	5542.
4' 6"	119.0	15.90	23'	3108.	415.5	86'	43450.	5809.
4' 7"	123.4	16.50	23' 6"	3245.	433.7	88'	45490.	6082.
4' 8"	128.0	17.10	24'	3384.	452.4	90'	47590.	6362.
4' 9"	132.6	17.72	24' 6"	3527.	471.4	92'	49720.	6648.
4' 10"	137.3	18.35	25'	3672.	490.9	94'	51920.	6940.
4' 11"	142.0	18.99	25' 6"	3820.	510.7	96'	54104.	7238.
5'	146.9	19.63	26'	3972.	530.9	98'	56420.	7543.
5' 3"	161.9	21.65	26' 6"	4126.	551.5	100'	58750.	7854.
5' 6"	177.7	23.76	27'	4283.	572.6			
5' 9"	194.3	25.97	27' 6"	4443.	594.0			

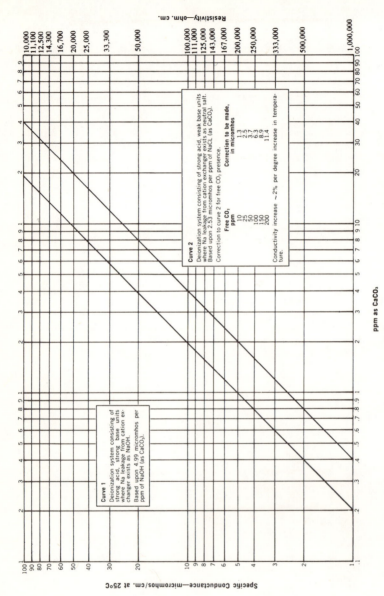

SPECIFIC CONDUCTANCE AND ELECTROLYTE CONTENT OF WATER VS. RESISTANCE AT 25° C

EFFECT OF MINERAL ACIDITY ON pH

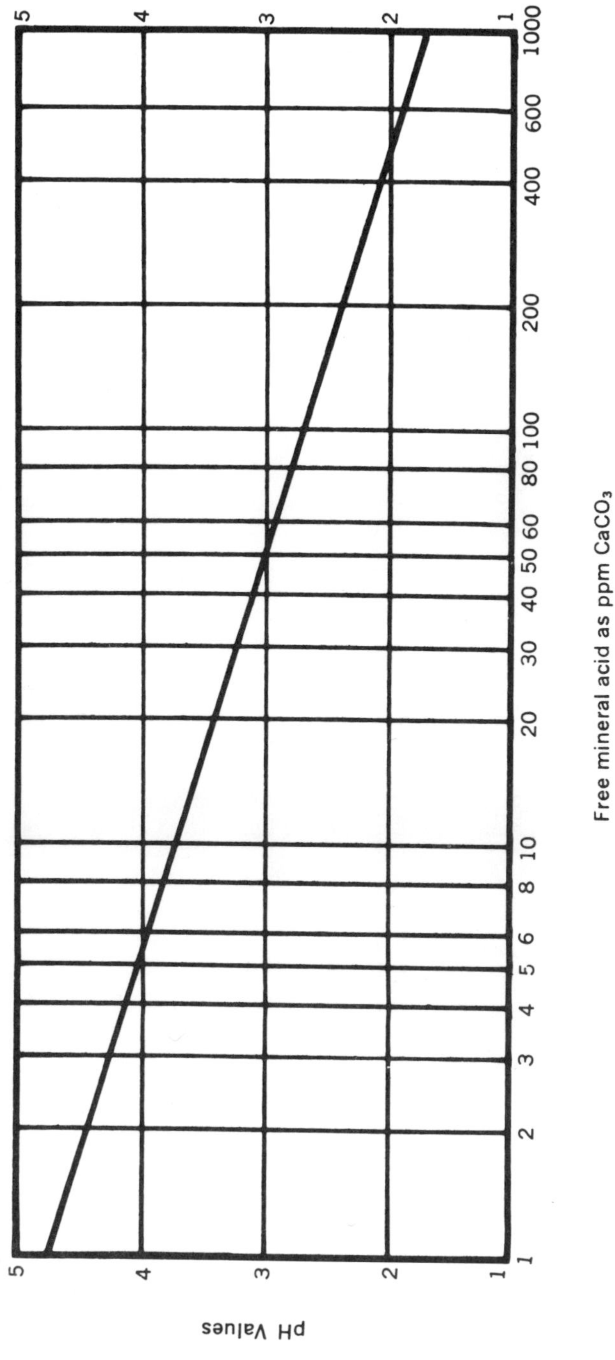

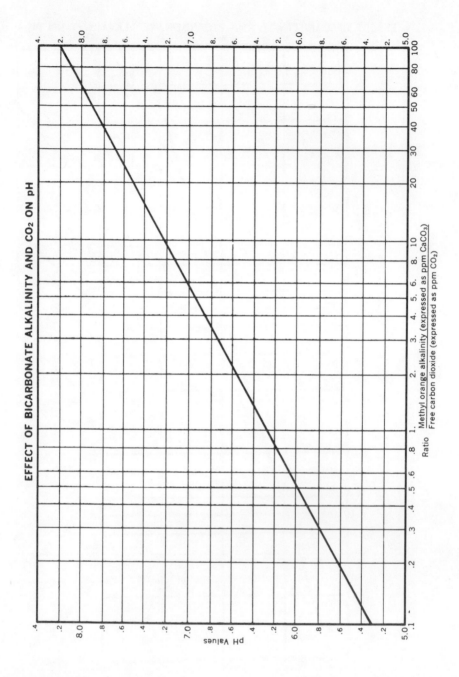

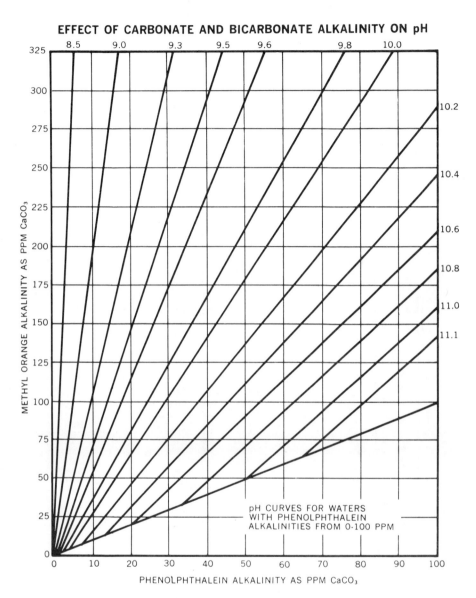

EFFECT OF CARBONATE AND BICARBONATE ALKALINITY ON pH

Note: pH value will also depend on temperature of water. Chart above is based on temperature of 20 to 25°C. As water temperature decreases, the pH value for any given combination of alkalinity forms will increase slightly above the value indicated on the chart. For example, at 5°C. actual pH will be about 0.2 units higher in 8.5 to 9.0 pH range; about 0.3 units higher in 9.0 to 10.0 pH range; and above pH 10 actual pH will be 0.4 to 0.6 pH units higher than indicated by chart.

FAHRENHEIT TO CENTIGRADE

−100 TO 0

C.	F.	
−73.3	−100	−148
−67.8	−90	−130
−62.2	−80	−112
−56.7	−70	−94
−51.1	−60	−76
−45.6	−50	−58
−40.0	−40	−40
−34.4	−30	−22
−28.9	−20	−4
−23.3	−10	14
−17.8	0	32

0 TO 100

C.		F.
−17.8	0	32
−17.2	1	33.8
−16.7	2	35.6
−16.1	3	37.4
−15.6	4	39.2
−15.0	5	41.0
−14.4	6	42.8
−13.9	7	44.6
−13.3	8	46.4
−12.8	9	48.2
−12.2	10	50.0
−11.7	11	51.8
−11.1	12	53.6
−10.6	13	55.4
−10.0	14	57.2
−9.44	15	59.0
−8.89	16	60.8
−8.33	17	62.6
−7.78	18	64.4
−7.22	19	66.2
−6.67	20	68.0
−6.11	21	69.8
−5.56	22	71.6
−5.00	23	73.4
−4.44	24	75.2
−3.89	25	77.0
−3.33	26	78.8
−2.78	27	80.6
−2.22	28	82.4
−1.67	29	84.2
−1.11	30	86.0
−0.56	31	87.8
0	32	89.6
0.56	33	91.4
1.11	34	93.2
1.67	35	95.0
2.22	36	96.8
2.78	37	98.6
3.33	38	100.4
3.89	39	102.2
4.44	40	104.0
5.00	41	105.8
5.56	42	107.6
6.11	43	109.4
6.67	44	111.2
7.22	45	113.0
7.78	46	114.8
8.33	47	116.6
8.89	48	118.4
9.44	49	120.2
10.0	50	122.0
10.6	51	123.8
11.1	52	125.6
11.7	53	127.4
12.2	54	129.2
12.8	55	131.0
13.3	56	132.8
13.9	57	134.6
14.4	58	136.4
15.0	59	138.2
15.6	60	140.0
16.1	61	141.8
16.7	62	143.6
17.2	63	145.4
17.8	64	147.2
18.3	65	149.0
18.9	66	150.8
19.4	67	152.6
20.0	68	154.4
20.6	69	156.2
21.1	70	158.0
21.7	71	159.8
22.2	72	161.6
22.8	73	163.4
23.3	74	165.2
23.9	75	167.0
24.4	76	168.8
25.0	77	170.6
25.6	78	172.4
26.1	79	174.2
26.7	80	176.0
27.2	81	177.8
27.8	82	179.6
28.3	83	181.4
28.9	84	183.2
29.4	85	185.0
30.0	86	186.8
30.6	87	188.6
31.1	88	190.4
31.7	89	192.2
32.2	90	194.0
32.8	91	195.8
33.3	92	197.6
33.9	93	199.4
34.4	94	201.2
35.0	95	203.0
35.6	96	204.8
36.1	97	206.6
36.7	98	208.4
37.2	99	210.2
37.8	100	212.0

100 TO 1000

C.		F.
38	100	212
43	110	230
49	120	248
54	130	266
60	140	284
66	150	302
71	160	320
77	170	338
82	180	356
88	190	374
93	200	392
99	210	410
100	212	413
104	220	428
110	230	446
116	240	464
121	250	482
127	260	500
132	270	518
138	280	536
143	290	554
149	300	572
154	310	590
160	320	608
166	330	626
171	340	644
177	350	662
182	360	680
188	370	698
193	380	716
199	390	734
204	400	752
210	410	770
216	420	788
221	430	806
227	440	824
232	450	842
238	460	860
243	470	878
249	480	896
254	490	914
260	500	932
266	510	950
271	520	968
277	530	986
282	540	1004
288	550	1022
293	560	1040
299	570	1058
304	580	1076
310	590	1094
316	600	1112
321	610	1130
327	620	1148
332	630	1166
338	640	1184
343	650	1202
349	660	1220
354	670	1238
360	680	1256
366	690	1274
371	700	1292
377	710	1310
382	720	1328
388	730	1346
393	740	1364
399	750	1382
404	760	1400
410	770	1418
416	780	1436
421	790	1454
427	800	1472
432	810	1490
438	820	1508
443	830	1526
449	840	1544
454	850	1562
460	860	1580
466	870	1598
471	880	1616
477	890	1634
482	900	1652
488	910	1670
493	920	1688
499	930	1706
504	940	1724
510	950	1742
516	960	1760
521	970	1778
527	980	1796
532	990	1814
538	1000	1832

SODIUM CHLORIDE SOLUTIONS (15°C. or 60°F.)

Baumé	Specific Gravity	Percent NaCl	Pounds per Gallon of Brine NaCl	Pounds per Gallon of Brine Water	Gallons-Water/ Gallons-Brine	Pounds NaCl/ Gallon-Water
0.6	1.004	0.528	0.044	8.318	0.999	0.044
1.1	1.007	1.056	0.089	8.297	0.996	0.089
1.6	1.011	1.584	0.133	8.287	0.995	0.134
2.1	1.015	2.112	0.178	8.275	0.993	0.179
2.7	1.019	2.640	1.224	8.262	0.992	0.226
3.3	1.023	3.167	1.270	8.250	0.990	0.273
3.7	1.026	3.695	1.316	8.229	0.988	0.320
4.2	1.030	4.223	1.362	8.216	0.987	0.367
4.8	1.034	4.751	1.409	8.202	0.985	0.415
5.3	1.038	5.279	1.456	8.188	0.983	0.464
5.8	1.042	5.807	1.503	8.175	0.982	0.512
6.4	1.046	6.335	0.552	8.159	0.980	0.563
6.9	1.050	6.863	0.600	8.144	0.978	0.614
7.4	1.054	7.391	0.649	8.129	0.976	0.665
7.9	1.058	7.919	0.698	8.113	0.974	0.716
8.5	1.062	8.446	0.747	8.097	0.972	0.768
9.0	1.066	8.974	0.797	8.081	0.970	0.821
9.5	1.070	9.502	0.847	8.064	0.968	0.875
10.0	1.074	10.030	0.897	8.047	0.966	0.928
10.5	1.078	10.558	0.948	8.030	0.964	0.983
11.0	1.082	11.086	0.999	8.012	0.962	1.039
11.5	1.086	11.614	1.050	7.994	0.960	1.094
12.0	1.090	12.142	1.102	7.976	0.958	1.151
12.5	1.094	12.670	1.154	7.957	0.955	1.208
12.9	1.098	13.198	1.207	7.937	0.953	1.266
13.4	1.102	13.725	1.260	7.918	0.951	1.325
13.9	1.106	14.253	1.313	7.898	0.948	1.385
14.4	1.110	14.781	1.366	7.878	0.946	1.444
14.8	1.114	15.309	1.420	7.858	0.943	1.505
15.3	1.118	15.837	1.475	7.836	0.941	1.568
15.8	1.122	16.365	1.529	7.815	0.938	1.629
16.2	1.126	16.893	1.584	7.794	0.936	1.692
16.7	1.130	17.421	1.639	7.772	0.933	1.756
17.2	1.135	17.949	1.697	7.755	0.931	1.822
17.7	1.139	18.477	1.753	7.733	0.929	1.888
18.1	1.143	19.004	1.809	7.710	0.926	1.954
18.6	1.147	19.532	1.866	7.686	0.923	2.022
19.1	1.152	20.060	1.925	7.669	0.921	2.091
19.6	1.156	20.588	1.982	7.645	0.918	2.159
20.0	1.160	21.116	2.040	7.620	0.915	2.229
20.4	1.164	21.644	2.098	7.596	0.912	2.300
21.0	1.169	22.172	2.158	7.577	0.910	2.372
21.4	1.173	22.700	2.218	7.551	0.907	2.446
21.9	1.178	23.228	2.279	7.531	0.904	2.520
22.0*	1.179	23.310	2.288	7.528	0.904	2.531
22.3	1.182	23.755	2.338	7.506	0.901	2.594
22.7	1.186	24.283	2.398	7.479	0.898	2.670
23.3	1.191	24.811	2.459	7.460	0.896	2.745
23.7	1.195	25.339	2.522	7.430	0.892	2.827
24.2	1.200	25.867	2.585	7.409	0.890	2.906
24.4	1.202	26.131	2.616	7.394	0.888	2.947
24.6	1.204	26.395	2.647	7.380	0.886	2.987

*Eutectic Point

SPECIFIC GRAVITIES $\left(\text{at } \frac{20°}{4°} \text{ C.}\right)$

AMMONIUM HYDROXIDE SOLUTION

Specific Gravity	Baumé	Per Cent NH_3	Normality	Grams per Liter	Lbs. per Cubic Foot	Lbs. per Gallon
0.9939	10.9	1	0.5836	9.939	0.6205	0.0829
0.9895	11.5	2	1.162	19.79	1.235	0.1652
0.9811	11.7	4	2.304	39.24	2.450	0.3275
0.9730	13.9	6	3.428	58.38	3.644	0.4872
0.9651	15.1	8	4.536	77.21	4.820	0.6443
0.9575	16.2	10	5.622	95.75	5.977	0.7991
0.9501	17.3	12	6.694	114.0	7.117	0.9515
0.9430	18.5	14	7.751	132.0	8.242	1.102
0.9362	19.5	16	8.796	149.8	9.351	1.250
0.9295	20.6	18	9.824	167.3	10.44	1.396
0.9229	21.7	20	10.84	184.6	11.52	1.540
0.9164	22.8	22	11.84	201.6	12.59	1.682
0.9101	23.8	24	12.82	218.4	13.64	1.823
0.9040	24.9	26	13.80	235.0	14.67	1.961
0.8980	25.9	28	14.76	251.4	15.70	2.098
0.8920	27.0	30	15.71	267.6	16.71	2.233

HYDROCHLORIC ACID (AQUEOUS HYDROCHLORIC ACID SOLUTIONS)

Specific Gravity	Baumé	Per Cent HCl	Normality	Grams per Liter	Lbs. per Cubic Foot	Lbs. per Gallon
1.0032	0.5	1	.2750	10.03	0.6263	0.0837
1.0082	1.2	2	.5528	20.16	1.259	0.1683
1.0181	2.6	4	1.117	40.72	2.542	0.3399
1.0279	3.9	6	1.691	61.67	3.850	0.5147
1.0376	5.3	8	2.276	83.01	5.182	0.6927
1.0474	6.6	10	2.871	104.7	6.539	0.8741
1.0574	7.9	12	3.480	126.9	7.921	1.059
1.0675	9.2	14	4.100	149.5	9.330	1.247
1.0776	10.4	16	4.728	172.4	10.76	1.439
1.0878	11.7	18	5.370	195.8	12.22	1.634
1.0980	12.9	20	6.022	219.6	13.71	1.833
1.1083	14.2	22	6.686	243.8	15.22	2.035
1.1187	15.4	24	7.363	268.5	16.76	2.241
1.1290	16.6	26	8.049	293.5	18.32	2.450
1.1392	17.7	28	8.748	319.0	19.91	2.662
1.1493	18.8	30	9.456	344.8	21.52	2.877
1.1593	19.9	32	10.17	371.0	23.16	3.096
1.1691	21.0	34	10.90	397.5	24.81	3.317
1.1789	22.0	36	11.64	424.4	26.49	3.542
1.1885	23.0	38	12.38	451.6	28.19	3.769
1.1980	24.0	40	13.14	479.2	29.92	3.999

SODIUM CARBONATE (AQUEOUS SOLUTIONS)

Specific Gravity	Baumé	Per Cent Na₂CO₃	Normality	Grams per Liter	Lbs. per Cubic Foot	Lbs. per Gallon
1.0086	1.2	1	0.1904	10.09	0.6296	0.0842
1.0190	2.7	2	0.3845	20.38	1.272	0.1701
1.0398	5.6	4	0.8979	47.59	2.596	0.3471
1.0606	8.3	6	1.201	63.64	3.973	0.5311
1.0816	10.9	8	1.633	86.53	5.402	0.7221
1.1029	13.5	10	2.081	110.3	6.885	0.9204
1.1244	16.0	12	2.545	134.9	8.423	1.126
1.1463	18.5	14	3.028	160.5	10.02	1.339

Specific Gravity	Baumé	Per Cent Na₂CO₃ +10H₂O	Normality	Grams per Liter	Lbs. per Cubic Foot	Lbs. per Gallon
1.0086	1.2	2.70	0.1904	27.23	1.700	0.2272
1.0190	2.7	5.40	0.3847	55.02	3.435	0.4592
1.0398	5.6	10.80	0.7853	112.3	7.010	0.9370
1.0606	8.3	16.20	1.2013	171.8	10.72	1.434
1.0816	10.9	21.60	1.6335	233.6	14.58	1.949
1.1029	13.5	27.00	2.0818	297.7	18.59	2.485
1.1244	16.0	32.40	2.5475	364.3	22.74	3.040
1.1463	18.5	37.80	3.0300	433.3	27.05	3.616

SODIUM HYDROXIDE

Specific Gravity	Baumé	Per Cent NaOH	Normality	Grams per Liter	Lbs. per Cubic Foot	Lbs. per Gallon
1.0095	1.4	1	0.2524	10.10	0.6302	0.0842
1.0207	2.9	2	0.5101	20.41	1.274	0.1704
1.0318	4.5	3	0.7814	30.95	1.932	0.2583
1.0428	6.0	4	1.042	41.71	2.604	0.3481
1.0538	7.4	5	1.317	52.69	3.289	0.4397
1.0648	8.8	6	1.597	63.89	3.988	0.5332
1.0758	10.2	7	1.902	75.31	4.701	0.6284
1.0869	11.6	8	2.175	86.95	5.428	0.7256
1.0979	12.9	9	2.470	98.81	6.168	0.8246
1.1089	14.2	10	3.772	110.9	6.923	0.9254
1.1309	16.8	12	3.392	135.7	8.472	1.133
1.1530	19.2	14	4.034	161.4	10.08	1.347
1.1751	21.6	16	4.699	188.0	11.74	1.569
1.1972	23.9	18	5.387	215.5	13.45	1.798
1.2191	26.1	20	6.094	243.8	15.22	2.035
1.2411	28.2	22	6.824	273.0	17.05	2.279
1.2629	30.2	24	7.577	303.1	18.92	2.529
1.2848	32.1	26	8.349	334.0	20.85	2.788
1.3064	34.0	28	9.145	365.8	22.84	3.053
1.3279	35.8	30	9.96	398.4	24.87	3.324
1.3490	37.5	32	10.79	431.7	26.95	3.602
1.3696	39.1	34	11.64	465.7	29.07	3.886
1.3900	40.7	36	12.51	500.4	31.24	4.176
1.4101	42.2	38	13.39	535.8	33.45	4.472
1.4300	43.6	40	14.30	572.0	35.71	4.773
1.4494	45.0	42	15.22	608.7	38.00	5.080
1.4685	46.3	44	16.15	646.1	40.34	5.392
1.4873	47.5	46	17.10	684.2	42.71	5.709
1.5065	48.8	48	18.08	723.1	45.14	6.035
1.5253	49.9	50	19.07	762.7	47.61	6.364

SULFURIC ACID (AQUEOUS SULFURIC ACID SOLUTIONS)

Specific Gravity	Baumé	Per Cent H_2SO_4	Normality	Grams per Liter	Lbs. per Cubic Foot	Lbs. per Gallon
1.0051	0.7	1	0.2051	10.05	0.6275	0.0839
1.0118	1.7	2	0.4127	20.24	1.263	0.1689
1.0184	2.6	3	.6234	30.55	1.907	0.2550
1.0250	3.5	4	.8360	41.00	2.560	0.3422
1.0317	4.5	5	1.053	51.59	3.220	0.4305
1.0385	5.4	6	1.271	62.31	3.890	0.5200
1.0453	6.3	7	1.493	73.17	4.568	0.6106
1.0522	7.2	8	1.717	84.18	5.255	0.7025
1.0591	8.1	9	1.945	95.32	5.950	0.7955
1.0661	9.0	10	2.174	106.6	6.655	0.8897
1.0731	9.9	11	2.408	118.0	7.369	0.9851
1.0802	10.8	12	2.643	129.6	8.092	1.082
1.0874	11.7	13	2.885	141.4	8.825	1.180
1.0947	12.5	14	3.126	153.3	9.567	1.279
1.1020	13.4	15	3.373	165.3	10.32	1.379
1.1094	14.3	16	3.619	177.5	11.08	1.481
1.1168	15.2	17	3.884	189.9	11.85	1.584
1.1243	16.0	18	4.127	202.4	12.63	1.689
1.1318	16.9	19	4.387	215.0	13.42	1.795
1.1394	17.7	20	4.647	227.9	14.23	1.902
1.1471	18.6	21	4.916	240.9	15.04	2.010
1.1548	19.4	22	5.181	254.1	15.86	2.120
1.1626	20.3	23	5.457	267.4	16.69	2.231
1.1704	21.1	24	5.728	280.9	17.54	2.344
1.1783	21.9	25	6.012	294.6	18.39	2.458
1.1862	22.8	26	6.289	308.4	19.25	2.574
1.1942	23.6	27	6.579	322.4	20.13	2.691
1.2023	24.4	28	6.864	336.6	21.02	2.809
1.2104	25.2	29	7.163	351.0	21.91	2.929
1.2185	26.0	30	7.455	365.6	22.82	3.051
1.2267	26.8	31	7.761	380.3	23.74	3.173
1.2349	27.6	32	8.059	395.2	24.67	3.298
1.2432	28.4	33	8.313	410.3	25.61	3.424
1.2515	29.1	34	8.676	425.5	26.56	3.551
1.2599	29.9	35	9.00	441.0	27.53	3.680
1.2684	30.7	36	9.311	456.6	28.51	3.811
1.2769	31.4	37	9.643	472.5	29.49	3.943
1.2855	32.2	38	9.961	488.5	30.49	4.077
1.2941	33.0	39	10.30	504.7	31.51	4.212
1.3028	33.7	40	10.63	521.1	32.53	4.349
1.3116	34.5	41	10.98	537.8	33.57	4.488
1.3205	35.2	42	11.31	554.6	34.62	4.628
1.3294	35.9	43	11.66	571.6	35.69	4.770
1.3384	36.7	44	12.01	588.9	36.76	4.914
1.3476	37.4	45	12.38	606.4	37.86	5.061
1.3569	38.1	46	12.73	624.2	38.97	5.209
1.3663	38.9	47	13.11	642.2	40.09	5.359
1.3758	39.6	48	13.47	660.4	41.23	5.511
1.3854	40.3	49	13.85	678.8	42.38	5.665
1.3951	41.1	50	14.22	697.6	43.55	5.821

SULFURIC ACID (AQUEOUS SULFURIC ACID SOLUTIONS) (CONTINUED)

Specific Gravity	Baumé	Per Cent H_2SO_4	Normality	Grams per Liter	Lbs. per Cubic Foot	Lbs. per Gallon
1.4049	41.8	51	14.62	716.5	44.73	5.979
1.4148	42.5	52	15.00	735.7	45.93	6.140
1.4248	43.2	53	15.41	755.1	47.14	6.302
1.4350	44.0	54	15.80	774.9	48.37	6.467
1.4453	44.7	55	16.22	794.9	49.62	6.634
1.4557	45.4	56	16.62	815.2	50.89	6.803
1.4662	46.1	57	17.05	835.7	52.17	6.974
1.4768	46.8	58	17.46	856.5	53.47	7.148
1.4875	47.5	59	17.91	877.6	54.79	7.324
1.4983	48.2	60	18.33	899.0	56.12	7.502
1.5091	48.9	61	18.79	920.6	57.47	7.682
1.5200	49.6	62	19.22	942.4	58.83	7.865
1.5310	50.3	63	19.68	964.5	60.21	8.049
1.5421	51.0	64	20.12	986.9	61.61	8.236
1.5533	51.7	65	20.61	1010	63.03	8.426
1.5646	52.3	66	21.06	1033	64.46	8.618
1.5760	53.0	67	21.55	1056	65.92	8.812
1.5874	53.7	68	22.00	1079	67.39	9.008
1.5989	54.3	69	22.51	1103	68.87	9.207
1.6105	55.0	70	22.98	1127	70.38	9.408
1.6221	55.6	71	23.51	1152	71.90	9.611
1.6338	56.3	72	23.98	1176	73.44	9.817
1.6456	56.9	73	24.51	1201	74.99	10.02
1.6574	57.5	74	25.00	1226	76.57	10.24
1.6692	58.1	75	25.55	1252	78.15	10.45
1.6810	58.7	76	26.06	1278	79.75	10.66
1.6927	59.3	77	26.59	1303	81.37	10.88
1.7043	59.9	78	27.10	1329	82.99	11.09
1.7158	60.5	79	27.65	1355	84.62	11.31
1.7272	61.1	80	28.18	1382	86.26	11.53
1.7383	61.6	81	28.73	1408	87.90	11.75
1.7491	62.1	82	29.24	1434	89.54	11.97
1.7594	62.6	83	29.79	1460	91.16	12.19
1.7693	63.0	84	30.30	1486	92.78	12.40
1.7786	63.5	85	30.85	1512	94.38	12.62
1.7872	63.9	86	31.34	1537	95.95	12.83
1.7951	64.2	87	31.87	1562	97.49	13.03
1.8022	64.5	88	32.34	1586	99.01	13.23
1.8087	64.8	89	32.85	1610	100.5	13.43
1.8144	65.1	90	33.30	1633	101.9	13.63
1.8195	65.3	91	33.79	1656	103.4	13.82
1.8240	65.5	92	34.22	1678	104.8	14.00
1.8279	65.7	93	34.64	1700	106.1	14.19
1.8312	65.8	94	35.09	1721	107.5	14.36
1.8337	65.9	95	35.55	1742	108.7	14.54
1.8355	66.0	96	35.93	1762	110.0	14.70
1.8364	66.0	97	36.34	1781	111.2	14.87
1.8361	66.0	98	36.68	1799	112.3	15.02
1.8342	65.9	99	37.36	1816	113.4	15.15
1.8305	65.8	100	37.34	1831	114.3	15.28

Tables and Conversion Factors

UNITS OF PRESSURE

Unit	Lbs. per Sq. in.	Feet of Water	Meters of Water	Inches of Mercury	Atmosphere	Kilograms per Sq. Cm.
1 Lb. per Sq. In. =	1	2.31	.704	2.04	.0681	.0703
1 Ft. of Water* =	.433	1	.305	.822	.02947	.0305
1 Meter of Water* =	1.421	3.28	1	2.89	.0967	.1
1 In. Mercury† =	.491	1.134	.3456	1	.0334	.0345
1 Atmosphere (at sea level) =	14.70	33.93	10.34	29.92	1	1.033
1 Kilogram per sq. Cm. =	14.22	32.8	10	28.96	.968	1

*Equivalent units are based on density of fresh water at 32° to 62° F.
†Equivalent units are based on density of mercury at 32° to 62° F.

UNITS OF POWER

Unit	Horse-power	Ft.-Lbs. per Minute	Watts	Kilo-watts	Metric Horse-power	B.T.U. per Minute
1 Horse-power =	1	33,000	746	.746	1.014	42.4
1 Ft.-Lb. per Minute =	.0000303	1	.0226	.0000226	.0000307	.001285
1 Watt =	.001340	44.2	1	.001	.001360	.0568
1 Kilowatt =	1.341	44,250	1000	1	1.360	56.8
(Cheval Vapeur) 1 Metric H.P. =	.986	32,550	736	.736	1	41.8
1 BTU per Min. =	.0236	778.4	17.6	.0176	.0239	1

UNITS OF FLOW

Units	U.S. gals. per minute	Million U. S. gal. per day	Cu. Ft. per second	Cu. meters per hr.	Liter per second
1. U.S. gal. per min. (U.S. G.P.M.)	1	.001440	.00223	.2270	.0631
1 million U.S. gals. per day (M.G.D.) =	694.5	1	1.547	157.73	43.8
1 cu. ft. per second =	448.8	.646	1	101.9	28.32
1 cu. meter per hour =	4.403	.00634	.00981	1	.2778
1 liter per second =	15.85	.0228	.0353	3.60	1

FLOW RATE*

	Bed Volumes/min.	Gal. (U.S.)/ cu. ft./min.	Gal. (Imp.)/ cu. ft./min.	Pounds H₂O/ cu. ft./min.
1 Bed Volume/min.	1	7.48	6.24	62.4
1 Gal. (U.S.)/cu. ft./min.	0.134	1	0.833	8.33
1 Gal. (Imp.)/cu. ft./min.	0.161	1.20	1	10
1 Pound H₂O/cu. ft./min.	0.016	0.12	0.10	1

*To convert flow rate per volume to flow rate per unit area, multiply flow rate per unit volume by resin volume and divide by cross-sectional area.

CAPACITY* AND REGENERATION LEVEL

	Meq./ml.	Pound equiv./ cu. ft.	Kilograins (as $CaCO_3$)/ cu. ft.	Grams CaO/ liter	Grams $CaCO_3$/ liter
1 Meq./ml.	1	0.0624	21.8	28	50
1 Pound equiv./cu. ft.	16.0	1	349	449	801
1 Kilograin (as $CaCO_3$)/cu. ft.	0.0459	0.00286	1	1.28	2.29
1 Gram CaO/liter	0.0357	0.00223	0.779	1	1.79
1 Gram $CaCO_3$/liter	0.0200	0.00125	0.436	0.560	1

*Capacity on a dry weight basis may be calculated as follows:

$$\text{gm.-meq./gm. of dry resin} = 6{,}240 \times \frac{\text{gm.-meq./ml.}}{\text{Wet density in lbs./cu. ft.} \times \% \text{ Solids}}$$

UNITS OF MASS

	Pounds	Grams	Kilograms	Grains	Kilograins
1 Pound	1	453.6	0.4536	7000	7
1 Gram	0.0022	1	0.001	15.43	0.01543
1 Kilogram	2.2	1000	1	15430	15.43
1 Grain	0.000143	0.065	0.000065	1	0.001
1 Kilograin	0.143	65	0.065	1000	1

UNITS OF DENSITY

	Pounds/Cubic ft.	Grams/liter	Pounds/Gal. (U.S.)	Pounds/Gal. (Imp.)
1 Pound/cu. ft.	1	16	0.134	0.160
1 Gram/liter	0.0624	1	0.00834	0.010
1 Pound/Gal. (U.S.)	7.48	120	1	1.2
1 Pound/Gal. (Imp.)	6.24	100	0.834	1

CONVERSION FACTORS FOR GALLONS-LITERS-CUBIC FEET

gal/min.	gal/hr.	gal/8hr.	gal/day	l/min.	l/hr.	l/8hr.	l/day	ft³/min.	ft³/hr.
1	60	480	1,440	3.78	227.12	1,817	5,451	0.1337	8.0209
5	300	2,400	7,200	18.93	1,135.6	9,085	27,255	0.6684	37.431
10	600	4,800	14,400	37.85	2,271.2	18,170	54,510	1.3368	80.209
15	900	7,200	21,600	56.78	3,406.9	27,255	81,765	2.0052	120.31
20	1,200	9,600	28,800	75.71	4,542.5	36,340	109,020	2.6736	160.42
25	1,500	12,000	36,000	94.63	5,678.1	45,425	136,274	3.3420	200.52
30	1,800	14,400	43,200	113.56	6,813.7	54,510	163,529	4.0104	240.63
35	2,100	16,800	50,400	132.50	7,949.3	63,595	190,784	4.6788	280.73
40	2,400	19,200	57,600	151.42	9,085.0	72,680	218,039	5.3472	320.83
45	2,700	21,600	64,800	170.34	10,221	81,765	245,294	6.0156	360.94
50	3,000	24,000	72,000	189.27	11,356	90,850	272,549	6.6840	401.04
60	3,600	28,800	86,400	227.12	13,627	109,020	327,059	8.0209	481.25
70	4,200	33,600	100,800	264.98	15,899	127,189	381,568	9.3577	561.46
80	4,800	38,400	115,200	302.83	18,170	145,359	436 078	10.694	641.67
90	5,400	43,200	129,600	340.69	20,441	163,529	490,588	12.031	721.88
100	6,000	48,000	144,000	378.5	22,712	181,700	545,098	13.368	802.09
110	6,600	52,800	158,400	416.4	24,984	199,869	599,607	14.705	882.29
120	7,200	57,600	172,800	454.2	27,255	218,039	654,117	16.042	962.50
130	7,800	62,400	187,200	492.1	29,526	236,209	708,627	17.379	1,042.7
140	8,400	67,200	201,600	530.0	31,797	254,379	763,137	18.715	1,122.9
150	9,000	73,000	216,000	567.8	34,069	276,334	817,646	20.052	1,203.1
160	9,600	76,800	230,400	605.7	36,340	290,719	872,156	21.389	1,283.3
170	10,200	81,600	244,800	643.5	38,611	308,889	926,666	22.726	1,363.5
180	10,800	86,400	259,200	681.4	40,882	327,059	981,176	24.063	1,443.8
190	11,400	91,200	273,600	719.2	43,154	345,228	1,035,686	25.399	1,524.0
200	12,000	96,000	288,000	757.1	45,425	363,398	1,090,195	26.736	1,604.2
220	13,200	105,600	316,800	832.8	49,967	399,738	1,199,215	29.410	1,764.6
240	14,400	115,200	345,600	908.5	54,510	436,078	1,308,234	32.083	1,925.0
260	15,600	124,800	374,400	984.2	59,052	472,418	1,417,254	34.757	2,085.4
280	16,800	134,400	403,200	1,060	63,595	508,758	1,526,173	37.431	2,245.8
300	18,000	144,000	432,000	1,136	68,137	545,098	1,635,293	40.104	2,406.3
320	19,200	153,600	460,800	1,211	72,680	581,437	1,744,312	42.778	2,566.7
340	20,400	163,200	489,500	1,287	77,222	617,777	1,853,332	45.452	2,727.1
360	21,600	172,800	518,400	1,363	81,765	654,117	1,962,351	48.125	2,887.5
380	22,800	182,400	547,200	1,438	86,307	690,457	2,071,371	50.799	3,047.9
400	24,000	192,000	576,000	1,514	90,850	726,797	2,180,390	53.472	3,208.3

WATER EQUIVALENTS

1 U.S. Gallon	=	.1337	Cubic Foot
1 U.S. Gallon	=	231.	Cubic Inches
1 U.S. Gallon	=	.833	British Imperial Gallon
1 U.S. Gallon	=	3.785	Liters
1 U.S. Gallon	=	3785.	Cubic Centimeters (Milliliters)
1 U.S. Gallon Water	=	8.33	Pounds (Lb.)
1 Cubic Foot	=	7.48	U.S. Gallons
1 Cubic Foot Water	=	62.43	Pounds (Lb.) (at greatest density—39.2°F)
1 Liter/Sec.	=	15.9	(U.S.) Gal./Min.
1 Cubic Meter/Hour	=	4.4	(U.S.) Gal./Min.
1 Kgr/Sq. Cm.	=	14.2	Pounds/Sq. Inch
1 Pound/1000 Gal.	=	120	Parts per Million
1 Inch/Minute Rise Rate	=	0.625	Gpm/Sq.Ft.
1 Cubic Meter	=	1000.	Liters
1 Cubic Meter	=	264.2	U.S. Gallons
1 Cubic Meter	=	220.	British Imperial Gallons
1 Cubic Meter	=	35.31	Cubic Feet
1 Boiler H.P. Hr.	=	4.	Gallons Water Evaporated per Hour

CONVERSION TABLE (Expressed to 3 Significant Figures)	Parts CaCO₃ per Million (ppm)	Parts CaCO₃ per Hundred Thousand (Pts./100,000)	Grains CaCO₃ per U.S. Gallon (gpg)	English Degrees or °Clark	French Degrees °French	German Degrees °German	Milli-equivalents per Liter or Equivalents per Million
1 Part per Million	1.0	0.1	0.0583	0.07	0.1	0.0560	0.020
1 Part per Hundred Thousand	10.0	1.0	0.583	0.7	1.0	0.560	0.20
1 Grain per U.S. Gallon	17.1	1.71	1.0	1.2	1.71	0.958	0.343
1 English or Clark Degree	14.3	1.43	0.833	1.0	1.43	0.800	0.286
1 French Degree	10.0	1.0	0.583	0.7	1.0	0.560	0.20
1 German Degree	17.9	1.79	1.04	1.24	1.79	1.0	0.37
1 Milli-equivalent per Liter or 1 Equivalent per Million	50.0	5.0	2.92	3.50		2.80	1.0

WATER ANALYSIS UNITS CONVERSION TABLE (Expressed to 3 Significant Figures)	Parts per Million (ppm)	Milligrams per Liter (mgm/L)	Grams per Liter (grms/L)	Parts per Hundred Thousand (Pts./100,000)	Grains U.S. Gallon (grs/U.S. gal)	Grains per British Imp. Gallon	Kilograins per Cubic Foot (Kgr/cu. ft.)
1 Part per Million	1.0	1.0	0.001	0.1	0.0583	0.07	0.0004
1 Milligram per Liter	1.0	1.0	0.001	0.1	0.0583	0.07	0.0004
1 Gram per Liter	1000.0	1000.0	1.0	100.0	58.3	70.0	0.436
1 Part per Hundred Thousand	10.0	10.0	0.01	1.0	0.583	0.7	0.00436
1 Grain per U.S. Gallon	17.1	17.1	0.017	1.71	1.0	1.2	0.0075
1 Grain per British Imp. Gallon	14.3	14.3	0.014	1.43	0.833	1.0	0.0062
1 Kilograin per Cubic Foot	2294.0	2294.0	2.294	229.4	134.0	161.0	1.0

Note: In practice, water analysis samples are measured by volume, not by weight and corrections for variations in specific gravity are practically never made. Therefore, parts per million are assumed to be the same as milligrams per liter and hence the above relationships are, for practical purposes, true.

SCREEN EQUIVALENTS

Sieve No.	U.S. Standard Opening mm.	U.S. Standard Opening Inches	Meshes per Inch	Tyler Standard Opening mm.	Tyler Standard Opening Inches	Meshes per Inch	British Standard Opening mm.	British Standard Opening Inches
12	1.68	0.0661	10	1.65	0.065	10	1.68	0.0660
14	1.41	0.0555	12	1.40	0.055	12	1.40	0.0553
16	1.19	0.0469	14	1.17	0.046	14	1.20	0.0474
18	1.00	0.0394	16	0.991	0.039	16	1.00	0.0395
20	0.84	0.0331	20	0.833	0.0328	18	0.853	0.0336
25	0.71	0.0280	24	0.701	0.0276	22	0.699	0.0275
30	0.59	0.0232	28	0.589	0.0232	25	0.599	0.0236
35	0.50	0.0197	32	0.495	0.0195	30	0.500	0.0197
40	0.42	0.0165	35	0.417	0.0164	36	0.422	0.0166
45	0.35	0.0138	42	0.351	0.0138	44	0.353	0.0139
50	0.297	0.0117	48	0.295	0.0116	52	0.295	0.0116
60	0.250	0.0098	60	0.246	0.0097	60	0.251	0.0099
70	0.210	0.0083	65	0.208	0.0082	72	0.211	0.0083
80	0.177	0.0070	80	0.175	0.0069	85	0.178	0.007
100	0.149	0.0059	100	0.147	0.0058	100	0.152	0.006
200	0.074	0.0029	200	0.074	0.0029	200	0.076	0.003
325	0.044	0.0017	325	0.043	0.0017	240	0.066	0.0026

COMMON CONVERSION FACTORS FOR ION EXCHANGE CALCULATIONS

To Convert from	to	Multiply by
Capacity		
Kgrs/ft.^{3} (as CaCO$_3$)	g CaO/l	1.28
Kgrs/ft.^{3} (as CaCO$_3$)	g CaCO$_3$/l	2.29
Kgrs/ft.^{3} (as CaCO$_3$)	eq/l	0.0458
g CaCO$_3$/l	Kgrs/ft^3 (as CaCO$_3$)	0.436
g CaO/l	Kgrs/ft^3 (as CaCO$_3$)	0.780
Flow Rate		
U.S. gpm/ft.^{3}	BV/hr	8.02
U.S. gpm/ft.^{2}	m/hr	2.45
U.S. gpm	m^3/hr	.227
BV/min	U.S..gpm/ft^3	7.46
Pressure Drop		
psi/ft.	mH$_2$O/m resin	2.30
	g/cm^2/m	230
Regenerant Concentration		
lbs/ft.^{3}	g/l	16.0
Density		
lbs/ft.^{3}	g/l	16.0
Rinse Requirements		
U.S. gal/ft.^{3}	BV	0.134

WEIGHTS AND MEASURES—Miscellaneous

Avoirdupois Weight
- 1 dram = 27 11/32 grains
- 1 ounce = 16 drams
- 1 ounce = 437½ grains
- 1 pound = 16 ounces
- 1 pound = 7000 grains
- 1 short ton = 2000 pounds
- 1 long ton = 2240 pounds
- NOTE: The grain is the same in avoirdupois, troy and apothecaries weight.

Troy Weight
- 1 pennyweight = 24 grains
- 1 ounce = 20 pennyweights
- 1 ounce = 480 grains
- 1 pound = 12 ounces
- 1 pound = 5760 grains

Apothecaries Weight
- 1 scruple = 20 grains
- 1 dram = 3 scruples
- 1 dram = 60 grains
- 1 ounce = 8 drams
- 1 ounce = 480 grains
- 1 pound = 12 ounces
- 1 pound = 5760 grains

Metric Weight
- 1 microgram = 0.001 milligram
- 1 milligram = 0.001 gram
- 1 centigram = 0.01 gram
- 1 decigram = 0.1 gram
- 1 dekagram = 10. grams
- 1 hectogram = 100. grams
- 1 kilogram = 1000. grams
- 1 metric ton = 1000. kilograms

Dry Measure (U.S.)
- 1 quart = 2 pints
- 1 peck = 8 quarts
- 1 bushel = 4 pecks
- 1 bushel = 32 quarts
- 1 barrel = 105 quarts

Wet Measure (U.S.)
- 1 pint = 4 gills
- 1 quart = 2 pints
- 1 gallon = 4 quarts
- 1 barrel (liquid) = 31½ gallons
- 1 barrel (oil) = 42 gallons

Long Measure
- 1 foot = 12 inches
- 1 yard = 3 feet
- 1 rod = 5½ yards
- 1 rod = 16½ feet
- 1 furlong = 40 rods
- 1 furlong = 660 feet
- 1 statute mile = 8 furlongs
- 1 statute mile = 5280 feet

Square Measure
- 1 square foot = 144 square inches
- 1 square yard = 9 square feet
- 1 square rod = 30¼ square yards
- 1 rood = 40 square rods
- 1 acre = 4 roods
- 1 acre = 43,560 square feet
- 1 square mile = 640 acres
- 1 square mile = 27,878,400 square feet

Cubic Measure
- 1 cubic foot = 1728.0 cubic inches
- 1 quart (liquid) = 57.75 cubic inches
- 1 gallon = 231.0 cubic inches
- 1 barrel (liquid) = 7276.5 cubic inches
- 1 cubic yard = 27.0 cubic feet
- 1 cord = 128.0 cubic feet
- 1 ton (shipping) = 40.0 cubic feet
- 1 quart (dry) = 67.2006 cubic inches
- 1 bushel = 2150.42 cubic inches
- 1 barrel (dry) = 7056.0 cubic inches

Metric Linear Measure
- 1 micron = 0.001 millimeter
- 1 millimeter = 0.001 meter
- 1 centimeter = 0.01 meter
- 1 decimeter = 0.1 meter
- 1 dekameter = 10. meters
- 1 hectometer = 100. meters
- 1 kilometer = 1000. meters

Metric Square Measure
- 1 sq. centimeter = 100 sq. millimeters
- 1 sq. decimeter = 100 sq. centimeters
- 1 square meter = 100 sq. decimeters
- 1 are = 100 square meters
- 1 hectare = 100 are
- 1 sq. kilometer = 100 hectares
- 1 square meter = 10,000 sq. centimeters
- 1 sq. kilometer = 1,000,000 square meters

Metric Cubic Measure
- 1 milliliter (cubic centimeter) = 1000 cubic millimeters
- 1 liter (cubic decimeter) = 1000 milliliters
- 1 hectoliter = 100 liters
- 1 cubic meter = 10 hectoliters
- 1 cubic meter = 1000 liters

Miscellaneous U.S. Liquid Measures — conversions to four significant figures
- 1 minum or drop = 0.0616 milliliters
- 1 fluid dram = 60 minums = 3.697 milliliters
- 1 fluid ounce = 8 fluid drams = 29.57 milliliters
- 1 quart = 32 fluid ounces = 0.9463 liters
- 1 gallon = 4 quarts = 3.785 liters
- 1 hogshead = 63 gallons = 2 barrels (liquid) = 0.2385 cubic meter
- 1 tun = 252 gallons = 8 barrels (liquid) = 0.9539 cubic meter

Linear Equivalents—to four significant figures

Inches	Feet	Yards	Miles	Millimeters	Centimeters	Meters	Kilometers
1.0	= 0.08333	= 0.02778	= 0.00001578	= 25.4	= 2.54	= 0.0254	= 0.0000254
12.0	= 1.0	= 0.3333	= 0.0001894	= 304.8	= 30.48	= 0.3048	= 0.0003048
36.0	= 3.0	= 1.0	= 0.0005682	= 914.4	= 91.44	= .9144	= 0.0009144
63360.0	= 5280.0	= 1760.0	= 1.0	= 1,609,000.0	= 160900.0	= 1609.0	= 1.609
0.3937	= 0.03281	= 0.01094	= 0.000006214	= 10.0	= 1.0	= 0.01	= 0.00001
39.37	= 3.281	= 1.094	= 0.0006214	= 1,000.0	= 100.0	= 1.0	= 0.001
39370.0	= 3281.0	= 1094.0	= 0.6214	= 1,000,000.0	= 100000.0	= 1000.0	= 1.0

Area Equivalents—to four significant figures

Square inches	Square feet	Square yards	Square centimeters	Square meters	Acres	Square miles	Ares	Hectares	Square kilometers
1.0	= 0.006944	= 0.0007716	= 6.452	= 0.0006452					
144.0	= 1.0	= 0.1111	= 929.0	= 0.0929					
1296.0	= 9.0	= 1.0	= 8361.0	= 0.8361					
0.155	= 0.001076	= 0.0001196	= 1.0	= 0.0001					
50.0	= 10.76	= 1.196	= 10000.0	= 1.0					
					1.0	= 0.001563	= 40.47	= 0.4047	= 0.004047
					640.0	= 1.0	= 25900.0	= 259.0	= 2.59
					0.02471	= 0.00003861	= 1.0	= 0.01	= 0.0001
					2.471	= 0.003861	= 100.0	= 1.0	= 0.01
					247.1	= 0.3861	= 10000.0	= 100.0	= 1.0

Volume Equivalents—to four significant figures

U.S. quarts	U.S. gallons	British imperial quarts	British imperial gallons	Cubic inches	Cubic feet	Liters	Cubic meters	Lbs. fresh water
1.0	= 0.25	= 0.8327	= 0.2082	= 57.75	= 0.03342	= 0.9464	= 0.0009464	= 2.089
4.0	= 1.0	= 3.331	= 0.8327	= 231.0	= 0.1337	= 3.785	= 0.003785	= 8.346
1.201	= 0.3002	= 1.0	= 0.25	= 69.35	= 0.04014	= 1.137	= 0.001137	= 2.506
4.804	= 1.201	= 4.0	= 1.0	= 277.4	= 0.1605	= 4.546	= 0.004546	= 10.02
0.01732	= 0.004329	= 0.01442	= 0.003606	= 1.0	= 0.0005787	= 0.01639	= 0.00001639	= 0.0381
29.92	= 7.481	= 24.92	= 6.229	= 1728.0	= 1.0	= 28.32	= 0.02832	= 62.43
1.057	= 0.2642	= 0.8799	= 0.22	= 61.02	= 0.03531	= 1.0	= 0.001	= 2.2046
1057.0	= 264.2	= 879.9	= 220.0	= 61020.0	= 35.31	= 1000.0	= 1.0	= 2204.
0.4787	= 0.120	= 0.4000	= 0.0998	= 27.7	= 0.0160	= 0.454	= .0004535	= 1.0

Weight Equivalents—to four significant figures

Grains	Avoirdupois ounces	Avoirdupois pounds	Grams	Kilograms
1.0	= 0.002286	= 0.0001429	= 0.06480	= 0.00006480
437.5	= 1.0	= 0.06250	= 28.35	= 0.02835
7000.0	= 16.0	= 1.0	= 453.6	= 0.4536
15.43	= 0.03527	= 0.002205	= 1.0	= 0.001
15430.0	= 35.27	= 2.205	= 1000.0	= 1.0

SUPPLEMENTAL READING

Textbooks

Applebaum, S., *Demineralization by Ion Exchange,* Academic Press, New York, NY, 1968.

Arden, P. V., *Water Purification by Ion Exchange,* Plenum Press, New York, NY, 1968.

Calmon, C., Kressman, T. R. E., *Ion Exchangers in Organic and Biochemistry,* Interscience, New York, NY, 1957.

Dorfner, K., *Ion Exchangers,* Ann Arbor Science Publishers, Inc., Ann Arbor, MI, 1972.

Helffrich, F., *Ion Exchange,* McGraw Hill, New York, NY, 1962.

Kunin, R., *Ion Exchange Resins,* Robert E. Krieger Publishing Co., Huntington, NY, 1972.

Marinsky, J., *Ion Exchange, Vol. 1, 2, 3, 5, and 6,* Marcel Dekker, New York, NY, 1966, 1969, and 1973.

Nachod, F., Schubert, J., *Ion Exchange Technology,* Academic Press, New York, NY, 1956.

Osborn, G. H., *Synthetic Ion Exchangers: Recent Developments in Theory and Application,* McMillan, New York, NY, 1958.

SUBJECT INDEX

A

ASME code, 34
Acetic acid, 28
Acid absorber, 70
   acetic, 28
   carbonic, 8, 17, 62, 66, 69, 106
   concentration, 83
   hydrochloric, 46, 77
   hypochlorous, 17
   organic, 53
   regeneration levels, 83
   silicic, 66, 69, 112
   strength, 28
   sulfuric, 28, 46, 77
   waste, 113
Acidity, free mineral (FMA), 7, 16, 82
   total mineral (TMA), 16
Adjustment of backwash flow rates, 18
Affinity for WAC resins, 62
Affinity of carboxylic exchange groups 63
Air, compressed, 119
   domes, 54
   flotation, 135
   mixing, 42, 102-104, 118, 119
Alkaline waters, 17
Alkalinity, 8, 15, 56, 60, 82, 127
Alkalinity end point, 65
   bicarbonate, 8
   carbonate, 8, 15
   hydrate, 8
   hydroxide, 15
Aluminum hydroxide, 133
American Water Works Assoc. (AWWA), 9
Amination, 28
Amines, filming, 134
Ammonia or Ammonium hydroxide, 28, 46

Analysis, 10, 11
   of organics in water, 20
   water, 9, 17
Anion exchange resins, 135
Anion resin — hydroxyl form, 128
Anion, layered, 113
   quality, 17, 46, 63, 96, 129, 130
Anions, 5
   total equivalent weight of, 12
   total mineral (TMA), 7
Anode, 5
Areas, dead, 96
Attack, oxidation, 67, 136, 140
Attraction of carboxylic group, 62
Attrition, mechanical, 135

B

Back pressure, 39, 71, 138
Back-regeneration, 65
Backwash, 26, 42, 71
   bed expansion, 36
   conditioning, 120
   cycle, 50, 70
   flow rate, 42
   water quality, 67
Balance of anions and cations, 10, 17
Barium, 9
Base strength to sodium hydroxide, 28
Bead, copolymer, 25
   resin, 25
   copolymer, 26
Bed, classified, 26
   expansion, 26, 50
Bicarbonate, 17, 66
   alkalinity, 8
Bisulfate, 94
Block bleed system, 46
Blocking flow of water, 105
Bond, double, 24

Bottom distributor, 42
Breakers, emulsion, 134
Breakthrough, 53, 68
Brine concentration, 74, 76
   regeneration, 103
   softening, 63
   waste, 147
Buffering effect, 53
Bulk chemical storage, 45

C

CO_2, free (carbon dioxide), 8
COD, 20
$CaSO_4$ precipitation, 80
Calcium carbonate, 63
Calcium sulfate, 51, 59, 63, 79, 83, 133
Calcium sulfate precipitation, 79, 83, 85, 86, 103, 108, 130
Calibration, 50
Capacities and efficiencies, 95
Capacity, 9, 10, 18, 75, 78, 82, 83, 87, 91, 93, 107, 118, 124-133
   curves, 84
   for divalent ions, 65
   for silica, 67
   loss, 51, 136
   maximum, 76
   of the WBA resins, 69
   operating, 128
   total, 73, 76
Capital costs, 103
Carbon dioxide, 17, 63, 66
   dissolved, 17
Carbonate, 66
   alkalinity, 8, 15
   calcium, 63
Carbonic acid, 8, 17, 62, 66, 69, 106, 171
Carboxylic, group, 62
   resins, 63, 85
Cathode, 5
Cation exchange resins, 134
   hydrogen form, 127
   quality, 17, 128, 134
Cations, 5

Caustic (sodium hydroxide), 46
   concentration, 88
   excess, 111, 112
   quality, 87
   strength to sodium hydroxide, 28
   waste, 112, 113
Changes, temperature, 134
Channel, 76
Channeling, 71, 75, 130-131
Characteristics, hydraulic, 26
Chemical precipitation, 66
Chemical regenerants, 146
Chlorate, sodium, 88
Chloride leakage, 69
Chlorine, 17
   removal, 17
Chloromethylation, 28
Classified bed, 26, 35, 67, 71, 117
Clumping, 129, 135
Co-current operation, 81, 101, 104
   systems, 57, 72
Coagulation and filtration, 19
Code, ASME, 34
Cold lime or lime-soda softening, 100
Cold lime with or without soda ash, 19
Colloidal contamination, 62, 127, 133
Compacted resin bed, 105
Compensation, temperature, 50, 51, 53
Complete deionization, 21
Complete water analysis, 9, 17
Compounds, organometallic, 133
   sequestering, 134
Compressed air, 119
Concentration, 3
   acid, 83
   brine, 74, 76
   caustic, 88
   of TDS, 30
   of hardness, 31
   of sodium ions, 31
   of sulfuric acid, 79
   of the regenerant, 81
   hydrochloric acid, 78, 85
   regenerant, 92
   sulfuric acid, 78, 86
Condensate polishing, 37, 114, 119, 120

Subject Index

Condition, dynamic, 30
  hydraulic, 37
  static, 57
Conditioning, backwash, 120
Conductivity, 53, 111, 114, 118
  meters, 51
Conserving water, 146
Constituents, minor, 9
Contact time, 75, 78, 85, 88-89, 91, 93
Contaminants, minor trace, 6
Contamination, colloidal, 133
  cross, 98, 114, 115, 117
  oil, 119
  organic, 70
Content, moisture, 135
Continuous ion exchange, 120
Continuous-process systems, 121
Control end point, 60
Controlling the flow for backwash, 44
Controls, level, 54
Conversion from ppm to epm, 12
Copolymer bead, 24-26
Correction factor, 123
Corrosion, 145
Cost penalties, 145
Costs, 67
  capital, 103
Countercurrent flow, 120
Countercurrent regeneration, 72, 101, 103, 104, 111, 112, 113, 114, 131
Cross-contamination, 98, 114, 115
Cross-linkage, 23, 25, 134
Curves, capacity, 84
Cycle, backwash, 50, 70
  regeneration, 72
  service, 57, 60, 61

D

DI product water, 96
  systems, 96
  water quality, 97
Damage, resin, 51
Dead areas, 96
Dead spots, 39, 42
Dealkalization, 66, 106-108
  SBA resin, 106
  split-stream, 106

Decarbonation, 48, 106
Decarbonator, forced-draft, 48
Decationized water, 95, 96, 106, 130, 131
Degasification (forced-draft or vacuum) 19, 46, 139
Degasifier, vacuum, 48, 139
Degassing, 66, 131, 113
Degassing system, 111
Dehydrate, 74
Dehydration, 75
Deionization, 9
Deionization, complete, 21
Deionized water, 4
Deterioration, resin, 17, 135
Diagram, flow, 139
Dilution water, 94, 95
  for caustic, 129
Dioxide, carbon, 17, 63, 66
Displacement rinse, 75, 95, 103, 105, 130, 146
Displacement rinse water, 90, 94, 95
Dissociated, 4
Dissolved carbon dioxide, 17
Dissolved solids, 7
Distillation, 19
Distilled water, 4
Distribution, 76
Distributor systems, 40-41, 129-130, 136
  and collectors, 38
  bottom, 42
  interface, 115, 117
  regenerant, 41
Divalent, 9, 12
Domes, air, 54
Double bond, 24
Drop, pressure, 35, 45, 71
Dynamic, 30
  conditions, 30
  flow, 57

E

EPM, 13
Earlier leakage of silica, 68
Eductors, 132, 138
Effect on the speed of reaction, 18

Effect, buffering, 53
Effective concentration, 32
Efficiency, regeneration, 65, 84, 92
Electric power, 140
Electrodes, 5
Electrodialysis, 19
Electroneutrality, 13
Emulsion breakers, 134
End point, 68
Epm or meq/l, 13
Equal valence, 12
Equal valence basis of weight, 12
Equilibrate, 138
Equilibrium, 4, 29, 30, 57, 73
Equilibrium point, 30
Equivalence per million, 13
Equivalent weight, 12, 13
Equivalent weight of, 60 66
Error, human, 138
Evaporator, 19
Excess caustic, 111, 112
Exchange, 62
   functional groups, 27, 85
   ion, 6
Expansion, bed, 26, 50

F

FMA 81, 89, 95, 112
Face piping and valves, 45
Factor, correction, 123
Fail-safe valves, 45
Failures, mechanical, 140
Fast or final rinse, 96, 104
Fast rinse flowrate, 19, 53, 95, 97
Faulty operation, 140
Filming amines, 134
Filter media, 70
Filtering capability of the mixed-bed, 119
Filtration, 19
Final or fast rinse, 103
Fines, resin, 67, 71, 120
Flocculant, 140
   fouling, 140
   polyelectrolyte, 70
Flotation, air, 135
   resin, 71, 119
Flow and capacity requirements, 36

Flow, countercurrent, 120
   diagram, 139
   dynamic, 57
   meters, 50
   per unit of resin volume, 30
   rate, 61, 67, 88, 92
   turbulent, 96
Flow- and temperature-sensitive, 65
Flowrate, linear, 126, 128, 129
Flowrates, regeneration, 81
Forced-draft decarbonator, 48, 106, 139
Form, sodium, 86
Foul the anion exchange resin, 48
Foulants, 19, 46, 64, 73, 87, 97, 128-130, 134, 140
   organic, 113, 138
Fouling, 20, 136
   flocculant, 140
   organic, 140
Four-bed system, 111
Free CO_2 (carbon dioxide), 8, 17
Free mineral acids (FMA), 7, 16, 82
Freeboard, 34, 130
Functional group, 27, 62

G

Gas solubility, 134
Gases, 48, 67, 128, 133
Gauges, pressure, 54
Gellular, 25, 134
Graded inert media underbedding, 44
Grain, term, 14
Grains per gallon as $CaCO_3$, 14, 15
Greensands, 29
Group, carboxylic, 62
   functional, 27, 62
   sulfonic, 27
Groups, exchange functional, 27

H

Hardness, 7, 9, 59, 65
   leakage, 60, 65, 76, 77, 101-104, 126
   load, 124, 125
   concentration, 31
   leakage, 63
   load, 124, 125

Hardness-to-alkalinity ratio, 108
Header-lateral, 40
Heads, strainer, 41
Heat of dilution, 129
High pressure drop, 135
  silica leakage, 67
  sodium leakage, 67
  temperature, 18, 67
  the alkalinity, 61
  viscosity, 18
Hot dilution water, 91, 96
Hot lime, 19
Hot lime-soda process, 100
Hub and spoke, 43
Human error, 128, 132, 138
Hydrate alkalinity, 8
Hydraulic characteristics, 1, 26, 37, 71, 134
  ram or water hammer, 45
  requirements, 126
Hydraulically classify, 117
Hydrochloric acid, 46, 77, 78, 85
Hydrogen-form SAC resins, 84
Hydrolysis, 98
  of chloride and sulfate, 115
  of sodium, 115
Hydrometers, 50
Hydroxide, alkalinity, 15
  aluminum, 133
  magnesium, 133
Hypochlorous acid, 17

I

Indicators, temperature, 50
Inert resin, 117
Inhibitors, scale, 134
Instructions, operating, 138
Instrumentation, 48, 130, 135, 136
Interface distributor, 115, 117
Interlock systems, 54
Intermediate base anion exchange resins, 69
Ion conversion factors, 14
Ion exchange, 6
  resin reactions, 29
  resin tanks, 32
Ions, 4
  balance of, 10
  negative, 5
  positive, 5
  ratios, 10, 56, 64, 78-79, 81-82, 88, 91-92
  total weight of, 11
Iron, 87, 133

J

Jackson turbidity units (JTU), 17

K

Kinetics, 1, 65

L

Layered anion, 112, 113
  on the top of the SAC resin, 108
Leaching of the regenerant, 96
Leakage, 67
  from the cation resins, 67
  chloride, 69
  hardness, 60, 63, 65, 76-77, 101-104, 126
  silica, 51-53, 67, 68, 88-89, 91, 111, 113, 128, 129, 137
  sodium, 15-18, 52, 64, 69, 80-84, 88-89, 108, 111, 113, 128, 137
Level, controls, 54
  regeneration, 76, 80, 81, 84, 87, 89, 93, 104
Life expectancy, 134, 135
Lime pretreatment, 17
Limit, pressure drop, 120
  solubility, 4
  temperature, 51, 91, 92
Linear flowrate, 126, 128, 129
Linkage, cross, 134
Load, hardness, 124, 125
Loss, capacity, 51, 136
  resin, 137
Lower limit, 30

M

Macroporous, 25, 134
Macroreticular, 25
Magnesium hydroxide, 133
Magnesium silicate, 146
Maintenance, 38, 128-130, 132-133, 135, 136, 138, 141, 144

Manpower changes, 132
Manual, operating, 139
Maximum, capacity, 76
　resin bed depth, 35
　vessel straight side, 35
Mechanical, attrition, 135
　failures, 140
Media, filter, 70
Membrane processes, 19, 146
Meters, conductivity, 51
　flow, 50
　pH 53
Micromhos, 52
Microorganisms, 128
Minimum, bed depth, 34, 113
　freeboard, 34
　pressure drop, 39
Minor constituents, 6, 9
Mixed-bed system, 42, 53, 54, 68
　separation, 18, 114
Mixing, air, 102, 104
　resins, 117, 119
Moisture content, 135
Monomers, 23, 27
Monovalent ions, 12
Multivalent ions, 9, 86, 127

N

Negative ions, 5
Nonreactive or colloidal silica, 66

O

Observations, 129
Ocean salinity, 3
Oil contamination, 119, 134
Open area, 39
Operating, capacity, 128
　instructions, 129, 132-133, 138
　manual, 139
Operation, co-current, 81, 101, 104
　countercurrent, 101, 103, 104
　faulty, 140
Optimum regeneration, 129
Organic acids, 53
Organic contamination, 70
Organic foulants, 113, 138, 140
　of SBA resins, 67, 68
Organics, 19, 64, 73, 127, 130, 134

Organometallic compounds, 133
Orifice plate, 105
Osmosis, 75
　reverse, 7, 19, 100
　shock, 75, 88, 89, 135
Oxidants, 17, 63, 128, 130, 134
Oxidation attack, 67, 98, 136, 140

P

Particulates, 17, 18, 66-67
pH 8, 17
　drops, 67
　meters, 53
Phenol formaldehyde resin structure, 21
Physical abuse of the resins, 121
Physical and hydraulic constraints, 34
Plate, orifice, 105
Point, end, 68
　equilibrium, 30
Polishers, mixed-bed, 68
Polishing, WAC resin, 112
　condensate, 37, 114, 119, 120
Polyelectrolyte flocculants, 70, 134
Polymer, 23
　cross-linked, 25
Polymeric silica, 66, 67, 88
Polystyrene, 24
Porosity, 25
Positive, block valves, 45
　charged ions, 5
　ions, 5
PPM as $CaCO_3$, 13-15
Precautions, safety, 78
Precipitates, 78, 80, 146
Precipitation, 51, 73, 78-79, 85, 95,
　108, 112-113, 130
　of calcium sulfate, 79, 80, 83, 85,
　86, 103, 108, 130
　chemical, 66
　of silica, 67, 113
　of sludges, 145
Preheating of the resin bed, 91
Pressure drop, 18, 26, 35, 45, 66, 70,
　71, 120, 130-131
Pressure gauges, 54
Pressure, back, 71, 138
Pretreatment, 10, 18-20, 134
　lime, 17

Subject Index

Preventive maintenance, 144
Problems, pressure drop, 70
  safety, 45
  scaling, 100
Process, membrane, 19, 146
  softening, 29
Pumps, 132, 138

Q

Quality, end point, 57
  of caustic, 87
  of produced water, 67, 72, 104
  of rinse water, 95
  of salt, 74
Quantities of salt, 31, 67
Quaternary ammonium ion exchange, 28

R

Rate, flow, 61, 67
Rates, reaction, 65
Ratio of silica to other anions, 67, 68
Ratios of ionic contaminants, 133
Reaction rates, 65
Reclamation of backwash water, 146
Record keeping, 133-135, 139
Recycle, 146
Recycle rinse, 97, 146
Recycling, 146
Reference test methods, 20
Regenerant, chemical, 31-32, 73, 129-130, 132, 140, 146
  concentration, 73-74, 78-79, 81, 92, 130-132
  distributors, 41
  flow rate, 42, 78-79, 87, 132
  level, 79, 93, 130
  purity, 73, 86, 130
  temperature, 78, 87, 90-91
Regeneration cycle, 72
  brine, 103
  countercurrent, 111, 112, 113, 114, 131
  distributor placement, 70
  efficiency, 65, 84, 92
  flowrates, 81
  high concentration, 31
  level, 76, 80-81, 84, 87, 89, 104

optimum, 129
steps, 57
stepwise, 130
time, 38, 126, 131
Rehydration, 75
Removal, of chlorine, 17
  of, silica, 18
Replacement, resin, 140
Requirements, hydraulic, 126
Resin, bead, 25
  cleaning, 19, 64, 73, 78, 129, 134
  damage, 51
  deterioration, 135
  fines, 71
  flotation, 71, 119
  interface, 35
  inert, 117
  level, 129
  life expectancy, 134
  loss, 67, 74, 119-120, 128-130, 137
  macroporous, 134
  mix, 118
  replacement, 140
  SAC (hydrogen form), 60, 77, 84, 106, 127, 134
  SAC (sodium form), 59, 74
  SBA (strongly basic anion), 65, 68, 87, 89, 92, 128, 135
  swell, 70
  tank linings, 137
  WAC (weakly acidic), 62, 108, 63
  WBA (weakly basic anion), 68, 70, 84-87, 92, 93
Resistance, 52, 53
Reverse Osmosis, 7, 19, 66, 100
Rinse, Displacement, 75, 95, 103, 130, 146
  recycle, 97, 146
  slow, 96
Run length, 72, 140

S

SAC 28
SAC + SBA — two-bed DI, 108
SAC + SBA DI systems, 127
SAC + WBA + SAC + SBA — four-bed DI, 113
SAC + WBA + SBA — three-bed DI, 111

SAC + WBA — two-bed DI, 108
SAC/SBA — mixed-bed DI, 114
SAC resin, (hydrogen form), 60, 77, 106
  (sodium form), 59, 74
  salt splitting, 28
SBA resin, 28, 65, 87-89
  dealkalization, 106
  type I, 68
  type II, 68
Safety precautions, 45, 78
Salinity, ocean, 3
Salt, 45
  quality, 74
Scale inhibitors, 134
Scaling problems, 100
Screen, well, 41
Screened distribution laterals, 34
Scrubber systems, 45
Seawater, 74, 76
Selection, ion exchange resins, 20
  screen size, 39
Selectivity, 59, 65, 87
  SAC resins, 62
Semiconductor, 114
Separation of, resins, 114
Sequestering compounds, 134
Series regeneration procedure, 112
Service cycle, 57, 60, 61, 104
Service flow rate, 42, 107
Service water temperature, 18, 48, 56-57, 67, 107
Shock, hydraulic, 44, 129, 131
  osmotic, 74, 75, 88, 89, 129, 134, 135
  thermal, 129, 134, 135
Silica (SiO_2), 12, 52, 66, 67, 87, 88, 89, 133
  capacity, 67
  leakage, 51, 52, 53, 67, 68, 88, 89, 91, 111, 113, 128, 129, 137
  precipitation, 113
  removal, 18
Silicate, magnesium, 146
Silicic acid, 66, 69, 112
Sites, exchange, 27, 85
Sizing of resin tanks, 34
Slow rinse, 96

Sludges, precipitation, 145
Soda ash (sodium carbonate), 46
Sodium, 67
  aluminum silicates, 29
  chlorate, 88
  chloride brine, 31
  concentration of ions, 31
  form, 86
  leakage, 15, 16, 18, 52, 60, 61, 64, 67, 69, 80-83, 88, 89, 108, 111, 113, 128, 137
  percent, 82
Soft water, 7, 65
Softener, systems, 124
  tanks, 33
Softening, high TDS water, 86
  process, 29
Softening, water, 9, 59, 63, 100
Solids, dissolved, 7
  suspended, 133
  total dissolved (TDS), 7, 11, 16
  total suspended, 7
Solubility, limit, 4
  gas, 134
Solutions, 79
Space velocity, 36, 37, 126, 128, 129
Spark test, 33
Specific ion monitoring, 53, 54
Specific resistance, 52, 53
Split neutral salts, 28
Split-stream dealkalization, 106
Spots, dead, 42
Stability, thermal, 68
Static, 30, 57
Steps, regeneration, 57
Stepwise regeneration, 130
Strainer heads, 41
Strong electrolyte resins, 93
Stronger affinity of calcium than sodium, 31
Strongly acidic cation resins, 28
Strongly basic anion resins, 28
Strontium, 9
Styrene-divinylbenzene resins, 24, 26
Sulfate, calcium, 63, 79, 83
Sulfonic group, 27
Sulfuric acid, 28, 46, 77
  concentration, 78, 79, 86

Subject Index

Superficial linear velocity, 36, 37
Supersaturated, 79
Support media, 39
Support of distributor laterals, 39
Support of screen-wrapped laterals, 39
Suspended solids, 133
Swelling and shrinkage, 63, 70
System, block bleed, 46
    co-current, 57, 72
    continuous-process, 121
    countercurrent, 72
    degassing, 111
    distributor, 136
    four-bed, 111
    interlock, 54
    mixed-bed, 42, 54
    scrubber, 45
    softener, 124

T

TDS, 10, 17, 30, 56, 107, 127, 145
TOC, 20
TSS, 127
Tank design pressures, 33, 128
Tanks, linings, 131
    softener, 33
Temperature, 18, 30, 61
    changes, 134
    compensation, 50, 51, 53
    effects on silica capacity, 18
    indicators, 50
    limitations, 51, 91-92
    near freezing, 62
    variations with seasons, 18
    water, 10, 18, 63, 71, 88, 92, 128, 130
Thermal, shock, 134-135
    stability, 68
Time, 30
    of the backwash, 71
    contact, 75, 78, 85, 88, 89, 91, 93
    regeneration, 38, 126, 131
Timers and switches, 55
Total, capacity, 73, 76
    dissolved solids (TDS), 7, 11, 16
    equivalent weight, 12
    mineral acidity (TMA), 16
    mineral anions (TMA), 7
    solids, 7
    suspended solids, 7
    weight of ions, 11
Training, 128-129, 135
Trivalent, 9, 12
Troubleshooting aids, 140
Turbidity, 17
Type II strongly basic anion resins, 28
Types of water used in backwash cycle, 71

U

Ultrafiltration, 19, 66
Underbedding, 34
Underbedding, upset, 135
Uniformity of water quality, 19
Unimixing, 102

V

Vacuum degasifier, 48, 139
Valences, 12
Valve leaks, 130
Valves, fail-safe, 45
    positive block, 45
Velocity, space, 36, 37, 126, 128, 129
Vents, 45
Viscosity change in water, 18
Volume of rinse water, 97

W

WAC + SAC + SBA — three-bed DI, 113
WAC resin (weakly acidic), 28, 62, 64, 70, 84-87, 92-93, 108
WBA resins (weakly basic anion), 68, 92
Waste, 78, 85, 103
    acid, 113
    alkaline, 17
    analysis, 9, 17
    brines, 147
    caustic, 112-113
    complete analysis, 17
    conserving, 146
    decationized, 95-96, 106, 129-131
    deionized, 4
    dilution, 95

hammer, 135, 137
pretreatment, 18
soft, 7
softening, 9, 59, 100
temperature, 71
viscosity goes down, 18
waste, 97
water, 97

Weak electrolyte resins, 93
Weakly acidic cation exchange resins, 27-28
Weakly basic anion exchange resins, 28
Weight, equivalent, 11-13
Well screen, 41

Z
Zeolites, 21